Post-harvest Tobacco Infestation Control

Post-harvest Tobacco Infestation Control

Edited by

L. Ryan

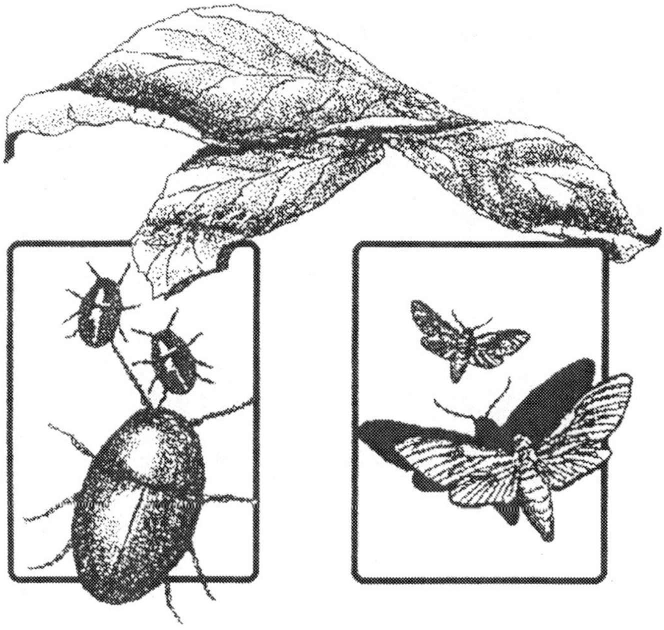

KLUWER ACADEMIC PUBLISHERS
DORDRECHT / BOSTON / LONDON

Library of Congress Cataloging-in-Publication Data

ISBN 0-412-63130-X

Published by Kluwer Academic Publishers,
P.O. Box 17, 3300 AA Dordrecht, The Netherlands.

Sold and distributed in North, Central and South America
by Kluwer Academic Publishers,
101 Philip Drive, Norwell, MA 02061, U.S.A.

In all other countries, sold and distributed
by Kluwer Academic Publishers,
P.O. Box 322, 3300 AH Dordrecht, The Netherlands.

Printed on acid-free paper

Reprinted 1999, 2001

Printed in the Netherlands

Contents

Acknowledgments

This manual was compiled by members of the Working Group on Infestation Control for Post-harvest Tobacco (ICPT). The working group and others authored and edited the different chapters. Members of the working group belong to cigarette manufacturing companies. Their affiliation is listed below:

Jane Carr, the late Mandy James, and Jeremy Ashworth (consultant) of Imperial Tobacco
Ronnie Compton, David Briggs and Paul Moore of Gallaher
Robert Genève of SEITA
Max Hutchens of R.J. Reynolds Tobacco
Eian David Massey, Terry Mitchell and Kord Westphal of British American Tobacco
Fred Moorcroft, Ed Cumming and Graham Dodd (consultant) of Rothmans
Wolfram Röper of Reemtsma
Lee Ryan of Philip Morris.

We would like to thank the following for their constructive comments on the draft manuscript: R. Carchman; D. Faustini; J.-M. Freymond; G. Koval, K. Milsom, B. Trochon (of CORALMA); and in particular Dr. Dennis Keever of the USDA for his detailed critique.
We are grateful to Terri Ryan for the cover design and some illustrations.
A big thank-you also to Veronique Prince for typing the manuscript.
The ICPT Working Group acknowledges the support of CORESTA.

Preface

In this manual, *Post-harvest Tobacco Infestation Control*, we have addressed the 'state-of-the-art' and given little account of obsolete techniques. With contributing authors from international cigarette manufacturers, plus consultation with the worldwide tobacco industry, we have recorded the acceptable methodology for infestation management. This manual fills a void, as the most recent treatment of this subject was more than 20 years ago.[350] Major emphasis is on **sanitation** which should, where possible, reduce or replace pesticide use at all stages of tobacco processing.

This manual is divided into an introduction and chapters dealing with: biology, monitoring, sanitation, physical control and insecticides – with separate chapters on insect growth regulators and fumigation. At the end, a few case histories are outlined to show how this integrated approach to infestation control is put into practice.

Comments from users of this manual regarding general usefulness, omissions and/or corrections are welcome and should be addressed to CORESTA, the infestation control subgroup of the Phytopathology group.

Chairman Wyden, Senator Hatch, and Members of the Committee, thank you for the invitation to testify on issues related to the status of the Highway Trust Fund and on options for financing highway improvements and construction.

SUMMARY

In 2013, governments at various levels spent $156 billion to build, operate, and maintain highways, and they spent $60 billion on mass transit systems. For both types of infrastructure, most of that spending was by state and local governments; about one-quarter of that total came from the federal government, mostly through the Highway Trust Fund. For several decades, the trust fund's balances were stable or growing, but more recently, annual spending for highways and transit has exceeded the amounts credited to the trust fund from taxes collected on gasoline, diesel fuel, and other transportation-related products and activities. Since 2008, in fact, lawmakers have transferred $54 billion from the U.S. Treasury's general fund to the Highway Trust Fund so that the trust fund's obligations could be met in a timely manner.

Moreover, with its current revenue sources, the Highway Trust Fund cannot support spending at the current rate. The Congressional Budget Office (CBO) estimates that, at the end of fiscal year 2014, the balance in the trust fund's highway account will fall to about $2 billion and the balance in its transit account will be only $1 billion. Spending for highways and transit will be $45 billion and $8 billion, respectively. By comparison, revenues collected for those purposes are projected to be $33 billion and $5 billion, respectively. The Department of Transportation (DOT) has indicated that it will probably need to delay payments to states at some point during the summer of 2014 in order to keep the fund's balance above zero, as required by law. Then, if nothing changes, the trust fund's balance will be insufficient to meet all of its obligations in fiscal year 2015, and it will incur steadily accumulating shortfalls in subsequent years. If lawmakers do not take action, all of the receipts credited to the fund in 2015 would be needed to meet obligations made before that year; none would be available to cover any new commitments that would be made in 2015.

Several options (or combinations of those options) could be pursued to address projected shortfalls in the Highway Trust Fund:

In: The Highway Trust Fund
Editor: Jonathan Caulfield

ISBN: 978-1-63482-012-7

Chapter 1

THE STATUS OF THE HIGHWAY TRUST FUND AND OPTIONS FOR FINANCING HIGHWAY SPENDING. TESTIMONY OF JOSEPH KILE, ASSISTANT DIRECTOR FOR MICROECONOMIC STUDIES, CONGRESSIONAL BUDGET OFFICE. HEARING ON "NEW ROUTES FOR FUNDING AND FINANCING HIGHWAYS AND TRANSIT"*

NOTES

Numbers in the text and tables may not add up to totals because of rounding.

Unless otherwise indicated, all years are federal fiscal years, which run from October 1 to September 30.

---

* This is an edited, reformatted and augmented version of testimony presented May 6, 2014 before the Senate Committee on Finance. The testimony was prepared by Sarah Puro and Chad Shirley with contributions from additional CBO staff.

Introduction 1

Tobacco is vulnerable to many insect pests while growing in the field.[4,313,357] Farmers may use pesticides to help control some insects and avoid losing up to 40% of the growing crop. Two insects, the cigarette beetle (*Lasioderma serricorne*) and the tobacco moth (*Ephestia elutella*), feed on cured tobacco leaves, whether air-cured burley, sun-cured oriental, flue-cured or tobacco by-products (Chapter 2). Cured tobacco is aged, usually for 1–3 years, to improve taste, aroma and quality characteristics, and during this time it is at risk of infestation. These two insect species and others will also infest tobacco seeds (section 2.3).

L. serricorne is *the* most serious insect threat to stored tobacco. Fortunately, methods to control this pest are effective against *E. elutella* also. Here, we will address the control of *L. serricorne* and mention *E. elutella* when its control is different than that of *L. serricorne*.

Worldwide yearly loss of stored tobacco to *L. serricorne* and *E. elutella* is estimated conservatively at 1%[326,350]: some $300 million. Imagine a warehouse with 5,000 tobacco containers – every year 50 of them are eaten by *L. serricorne*. This would represent a $100,000 loss, and may cause consumer complaints. Because *L. serricorne* larvae eat tobacco by-products, they become established within processing facilities in tobacco dust accumulations. These in-house infestations contaminate finished cut filler tobacco (rag) and may cause consumer complaints. Damage to tobacco is caused primarily by the insect larvae[173,183,243,275,350], although adult beetles may feed to a limited extent.[242,243] The larvae not only eat the tobacco but also leave excreta, cause cigarette spotting from body oils, and make holes in the leaves and cigarettes (see sections 9.1 and 9.7).

Constant vigilance and a comprehensive, integrated pest management programme is required to ensure a *quality* product without live insects or dead insect bodies, and to prevent the loss of millions of dollars of tobacco while in storage. An effective pest management programme requires a team approach involving active participation from top management. The role of the sanitarian, quality assurance personnel and pest control technician is to understand how the biology of the pest relates to the tobacco process. Then, a comprehensive monitoring programme during the tobacco process from the farm to the retailer (Fig. 1.1) helps determine how to effectively manage infestations. Control is best achieved by *prevention.* Failing this, a number of simple methods may be integrated to eliminate infestations.

Prevention is essential and should include segregation of different types of tobacco and stock rotation. **Good housekeeping practices or sanitation cannot be over-emphasized.** Wherever tobacco is conditioned, the cycle should achieve *L. serricorne* and *E. elutella* eradication in the tobacco entering the factory. Only when this control strategy is not correctly implemented, and therefore fails, are pesticides necessary (Table 1.1).

Do not buy infested tobacco. If you find that your tobacco has become infested after purchase fumigate it before putting it with your "clean" stocks. Remember, a few infested tobacco packages can infest your whole warehouse and/or factory. No matter how good your pest control programme, you must depend on your suppliers for pest-free tobacco. Encourage them to adopt the good manufacturing practices outlined here. If you put clean incoming tobacco into clean warehouses and clean

Table 1.1 The flow of tobacco from the farm to retail outlets (Fig. 1.1) with the possible IPM tactic for each stage

Tobacco process	Monitoring	Sanitation	Heat	Fumigation	Kabat®	Dianex®
1. On-farm storage	X	X				
2. Auction sales floor	X	X				
3. Stemmery manipulation	X	X	X		X	X
4. Warehouse	X	X		X		
5. Manufacturing facility	X	X	X			X
6. Finished goods warehouse	X	X				
7. Retail outlets	X	X				
8. In transit between stages	X	X				

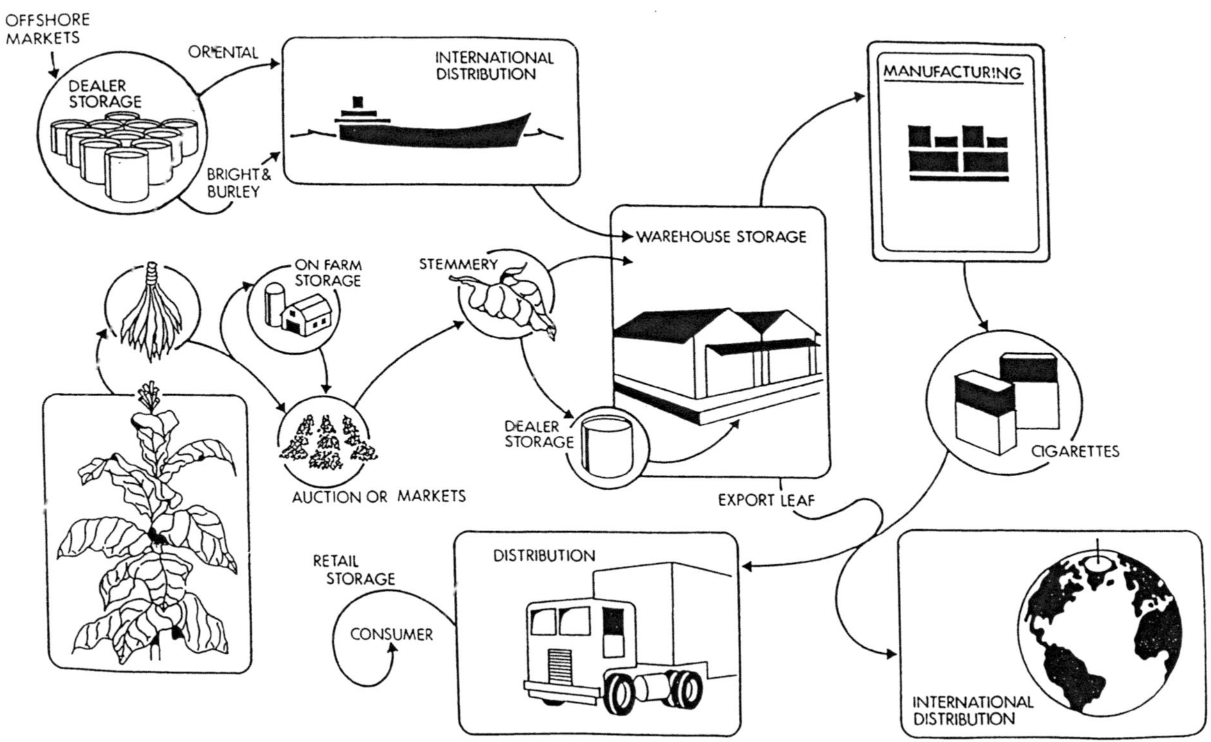

Fig. 1.1 The flow of tobacco from the farm to consumer via: on-farm storage, auctions or markets, stemmery, storage and ageing, manufacturing, distribution.

processing facilities, you will reduce potential *L. serricorne* and *E. elutella* infestations. Indeed, if a warehouse *is* infested, stop taking tobacco out until the tobacco is fumigated and pest-free.

The best *L. serricorne* and *E. elutella* control in storage complexes is achieved by monitoring incoming shipments and fumigating them prior to receipt for storage, when necessary. The best *L. serricorne* control in processing facilities is achieved by excellent housekeeping. Controlling *L. serricorne* to the "no problem" level requires a concerted effort on monitoring, prevention, housekeeping and judicious applications of methoprene or pyrethrum and fumigation. Methoprene (Kabat®) may greatly reduce the need for fumigations, but good sanitation will **always** be essential. As infestation problems disappear, so your efforts and expenses decline, but remember, *L. serricorne* is everywhere. You always must be vigilant and practice preventive measures.

The following chapters explain in detail each facet of an integrated pest management (IPM) programme for processed tobacco. We provide choices for tailoring the IPM programme to the different circumstances around the world. The key to success is **sanitation** by **vacuum** cleaning, but there are numerous routes to success (Chapter 4). Fumigation of infested tobacco (Chapter 8) will usually involve phosphine but other fumigants are available. When sanitation and fumigation are not effective we may resort to space spray or crack and crevice application of insecticides (Chapter 6). Increasing use is being made of temperature treatments (Chapter 5) and methoprene (Chapter 7). So, we have choices for flexibility, but all our decisions should be dictated by careful insect monitoring (Chapter 3). All aspects of the IPM should be: written as procedures, allowing for verification; reviewed, improved and focused, as circumstances change; audited by supervisory/consultant personnel.

We feel that the information contained in the following pages is a status report of IPM programmes for processed tobacco. This will serve as a menu for fellow-workers to choose facets and integrate them into their own IPM.

Biology 2

2.1 *LASIODERMA SERRICORNE* – THE CIGARETTE BEETLE

The cigarette beetle, *Lasioderma serricorne* (Fabricius), (Coleoptera: Anobiidae), is the major pest of cured tobacco, infesting this commodity during storage, manufacture, and at retail outlets.[350] *L. serricorne* were found in the tomb of Tutankhamun[7] and Rameses II.[327] Thus they were present in the Old World over 3,000 years ago, and were probably introduced from there to the New World, not *vice versa.*[297] Morphologically, the beetle has changed little in the interim.[286] *L. serricorne* was described from North America by Fabricius in 1792.[275] First records of its association with tobacco were made in Paris in 1848[297] and in America in 1886.[339] *L. serricorne* is now distributed worldwide, being of considerable adverse economic importance in tropical and temperate climates.[350] A recent review[17] was derived from this chapter and appears here with kind permission from Elsevier Science Ltd, UK.

2.1.1 MORPHOLOGY AND LIFE HISTORY

Adults

Adult beetles are 2.0–3.7 mm long,[289] weigh 1.6 to 4.4 mg and are a light to dark brown colour,[200,350] although a black body colour mutation is known.[100] Size depends on the type of food, [183] temperature, and humidity encountered during development.[350] Females tend to be larger than males.[215,218] The adult beetle (Plate 2.1) is oval, with the first thoracic segment (prothorax) bent down, and the head deflexed and obscured from above, giving the insect a humped, convex appearance of a burrowing insect. The wing covers (elytra) are not striated and appear smooth, unlike those of the drugstore beetle, *Stegobium paniceum* (Linnaeus).[173,200] The antennae (Plate 2.2) are saw-like (serrate), the same thickness from base to tip, and differ from the typical Anobiid antenna (as in *S. paniceum*) which form a broad three-segmented club.[162,173,200] Indeed, these and other features such as tarsal formula (number of segments in the "feet") and mandibular structure (front pair of mouthparts)[200] are used in keys to identify *L. serricorne.* [55,116,173,183,273]

Sex determination is performed by applying light pressure to the abdomen to evert the simple, transparent, female oviduct or complex sclerotized male claspers.

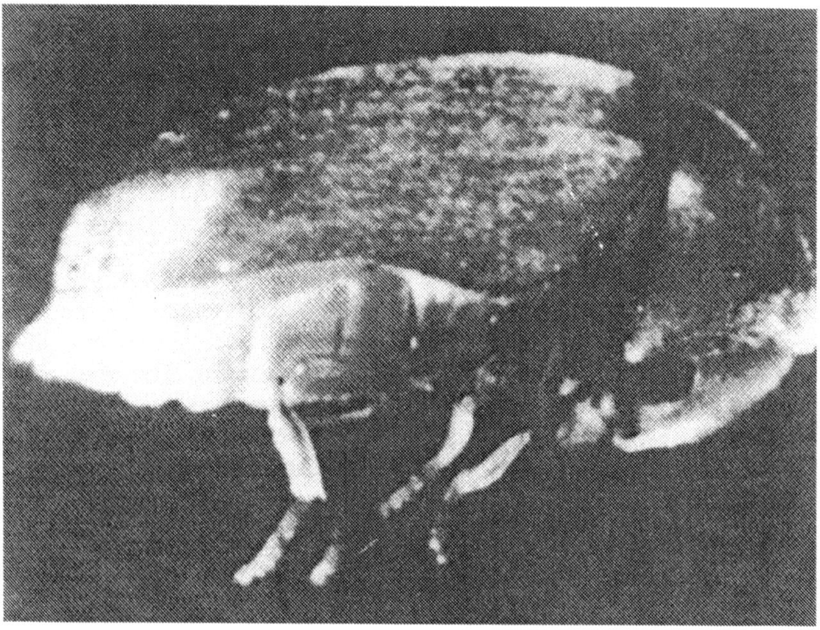

Plate 2.1 Adult *L. serricorne*.

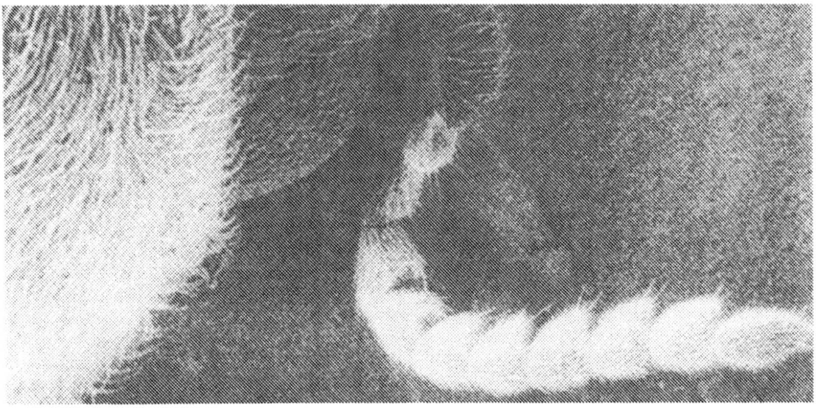

Plate 2.2 The serrate antenna of *L. serricorne*.

Adults have up to a 3 km range of flight,[62] which explains the local spread of an infestation.[173] They fly slowly,[289] as temperature permits,[173] most commonly during the late afternoon, dusk and during darkness.[173,287] They sometimes fly in daylight on warm, dull days particularly when egg laying.[18,69,352] Egg laying (oviposition) is stimulated by the chemical[139] and physical[202] nature of the substrate and deterred by the pheromone serricorole.[177]

Eggs

Eggs are oviposited onto dried material beginning one day after the sexually mature female emerges from the pupal cell and mates.[105] The eggs are pearly white,[289] 0.4 to 0.5 mm long and 0.2 mm wide.[183,318] The eggs weigh about 8 μg and, although fragile, have a waxy shell to protect them from desiccation. The eggs hatch (Plate 2.3) in 6 to 19 days[289,301] and the egg shell is eaten by the newly emerged larva,[265] which emerges apically from the area covered by spines.[302,318] Methods are available for collecting eggs from *L. serricorne* laboratory cultures.[135,201]

Larvae

Larvae (Plate 2.4) complete four growth stages (instars),[260] growing from less than 1.0 mm to 4.5 mm long and from 2.5 to 5.0 mg as fourth instars. The first three instars are of equal duration, with the fourth being twice as long; mortality is highest in the first instar.[173] Larvae are creamy coloured or greyish white and covered with fine light brown hairs.[318] Newly hatched larvae move away from light[56] and are extremely active, entering small holes in search of food and penetrating packaged commodities such as cigarette packs via the pack seal.[297] They cause damage by eating

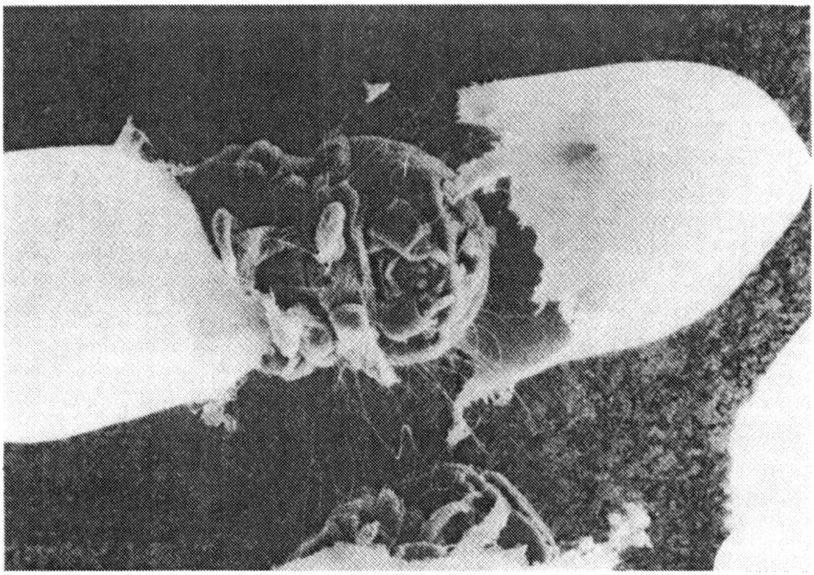

Plate 2.3 A larval *L. serricorne* emerging from the egg.

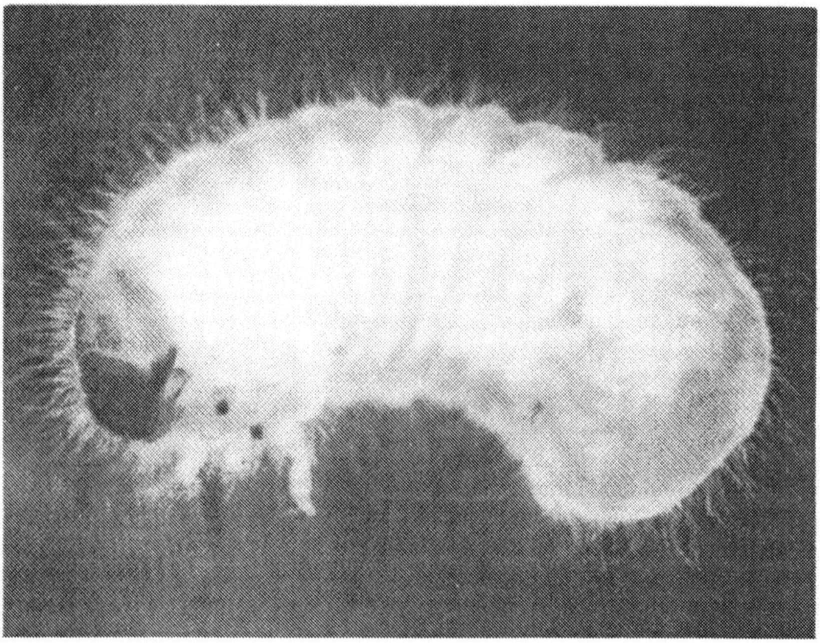

Plate 2.4 A full grown *L. serricorne* larva with a darker, sclerotized, head and numerous body hairs.

13–16 mg of the tobacco as they wander in the tobacco mass.[173,210,217,218,318] Penetration may be deep into loosely packed commodities, but they remain in the peripheral 50 mm in tightly packed tobacco strips.[216,318,350] Activity ceases when the temperature falls below 15°C[173,297] and the beetle overwinters in the larval stage.

Pupae

Pupae (Plate 2.5) develop in the pupal cell within the tobacco or attached to a surface, such as tobacco processing equipment, by biological cement. The cell (Plate 2.6) is made by the larva of food and excreta cemented together by a secretion produced by the midgut and perhaps cephalic glands.[173,354] The larva moults after 2 to 4 days (at 30°C) in the completed cell to become the pupa. The pupal period varies between 4 to 12 days[173,297,301] when it moults to become the adult. Adults remain in the cell to harden and become sexually mature.[173] Emergence from the pupal cell is mainly during the night.[301] Sex determination (Plate 2.7) of the pupae is by examination of the genital papillae which are globular and not projecting in the males, and three-segmented and divergent in the females.[161] The sex ratio of the emerged population is 1:1.[318]

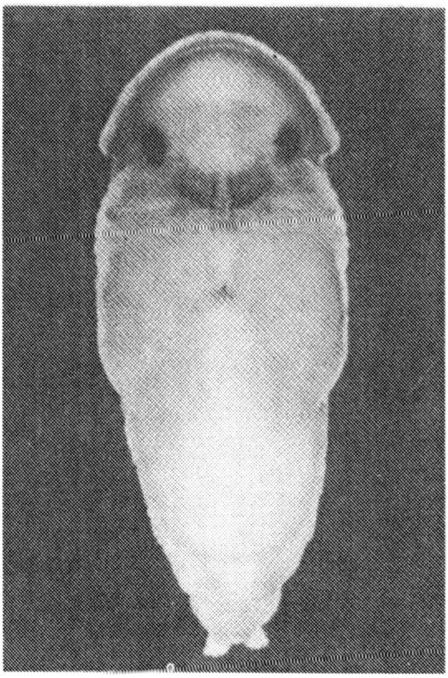

Plate 2.5 *L. serricorne* pupa removed from the pupal case.

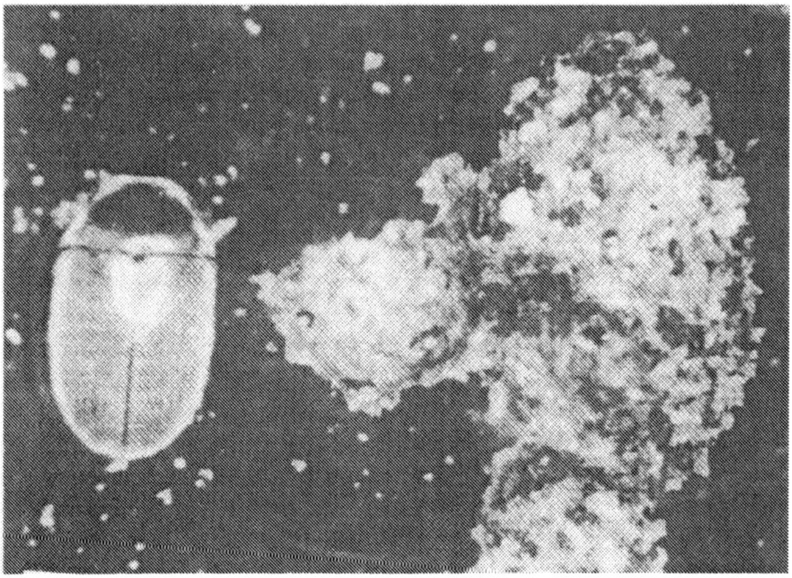

Plate 2.6 An adult *L. serricorne* and its pupal case made from frass and tobacco.

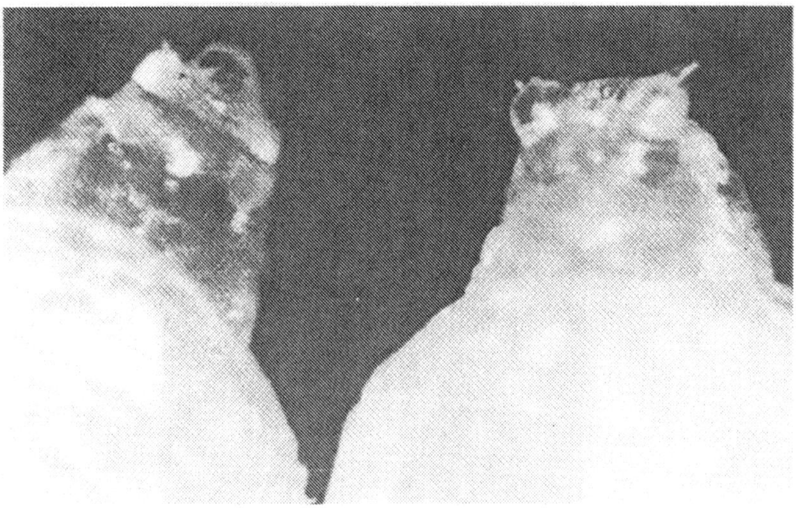

Plate 2.7 The terminal pupal segment (see Plate 2.5) of *L. serricorne*, to show the difference between male (left) and female.

2.1.2 SURVIVAL AND DEVELOPMENT

The data presented in this section relate to *L. serricorne* reared in the laboratory on artificial diets. The little information available shows that *L. serricorne* reared on tobacco, in the laboratory, have a mean egg to adult period of 66 days.[136,297] It is important to remember this difference between the laboratory and the field.

Egg to adult development, on artificial diets (Table 2.1), takes about 40 days,[136,173,275,289,301,318] and is dependent on food, temperature and humidity.[113,196] Of course, this may be considerably longer when the larvae overwinter.[241] The duration of the developmental period has been related to temperature and relative humidity[124,260,275] (Figs 2.1 and 2.2).

Populations increase quickly, a protected breeding pair produced 2,000 offspring via subsequent generations in 4 months[173] and have the capacity to multiply by 4 to 6 times its original size every 3 weeks.[215] This intrinsic rate of increase (*r*) has been related to temperature and humidity via the wet-bulb temperature.[113] Optimal conditions for rearing *L. serricorne* are 28°C and 70% RH[62,136,140,173,275] on 20:1 flour: yeast mixtures[140,301,316] or ground tobacco,[136] although, more precise or complicated diets have been published.[8,185,267,316] The adult lives 2 to 7 weeks,[173] again depending on larval food, temperature and humidity, life being shortened by high temperature and low humidity. Indeed, a temperature of 40°C for 24 hours is fatal to all life stages of *L. serricorne* (see section 5.1.2).

Oviposition increases from 15°C to optimal at 27°C.[140,173] Eggs develop and hatch between 20°C and 34°C, while larvae develop between 20°C and 37°C

Table 2.1 Duration of life stages of *L. serricorne* at 30 ± 2°C, 78% RH, fed on 10:1 wheat flour: baker's yeast[301]

	Developmental duration (days)	
Stage	Range	Mean ± SD
Egg	10–19	15 ± 3
Larva	10–18	13 ± 3
Pupa	4–12	8 ± 3
Adult	16–19	17 ± 1
Egg to adult	42–45	44 ± 1

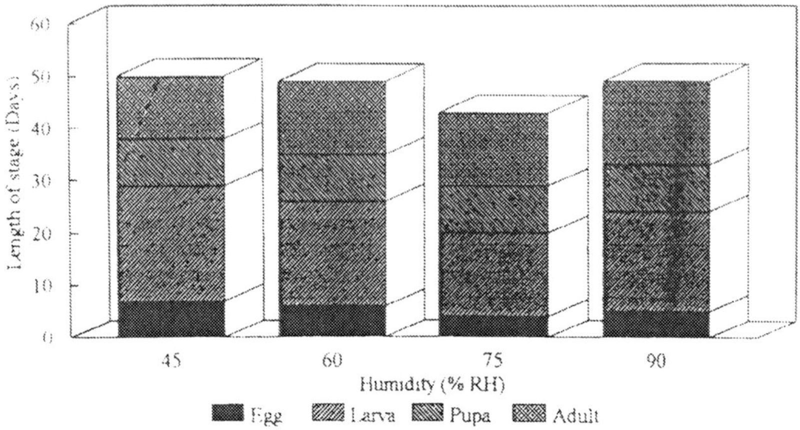

Fig. 2.1 The effect of humidity on the length of each separate life stage of *L. serricorne* at 32°C.[275]

at various humidities.[173] All stages of *L. serricorne* may exist at temperatures between 2°C and 36°C, and may survive below 2°C for short periods.[275]

In the warehouse, temperature and humidity gradients may form after a period of cooling, followed by an influx of warmer air that raises the temperature around the tobacco. Larvae prefer lower humidities when offered a gradient and prefer temperatures between 25°C and 28°C.[142] Temperature and relative humidity gradients may cause larvae to emigrate from within the tobacco to the surface, fall to the floor and die

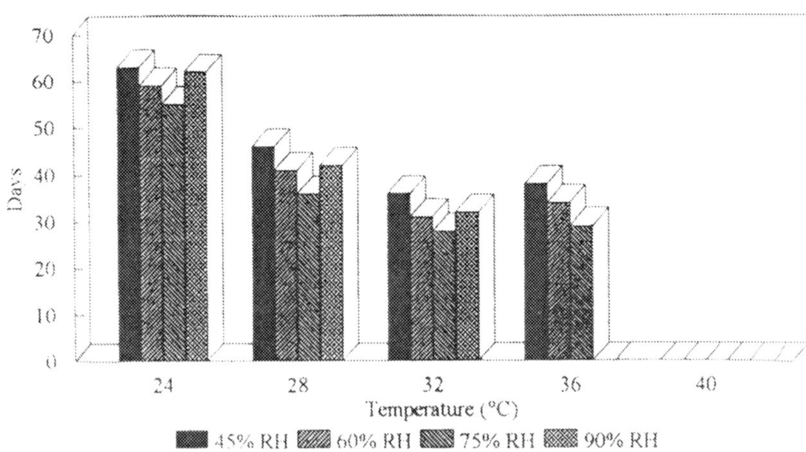

Fig. 2.2 The effect of temperature on the length of the life cycle of *L. serricorne* at different relative humidities (%RH) on a yeast diet.

from desiccation and exposure.[142] The pupa is relatively unaffected by humidity.[173]

2.1.3 TROPHISM

L. serricorne infests a wide range of foodstuffs including ground grain or seed, whole seeds, grains, rice, pulses, beans, pasta, spices, dried fruit, dried vegetables, yeast, dried insects, dried fish, fishmeal, meatmeal, leather, wax, and dried and processed plant material such as tobacco.[9,146,173,373] Transfer of a population from one foodstuff to another causes a variation in performance or fitness.[136]

When feeding on tobacco, *L. serricorne* survives best on high sugar/low nicotine tobacco, so flue-cured and oriental are most at risk.[185,203,242,295] Females are attracted to tobacco[203] and will lay more eggs on better grade flue-cured.[209] *L. serricorne* is nicotine-tolerant up to 4%.[242,370] It metabolizes, and hence detoxifies, nicotine into cotinine and other alkaloids.[307] The larvae are unable to survive on a tobacco diet containing 8.25% nicotine.[185,242]

Food requirements

L. serricorne has yeast-like intracellular symbionts in the gut cecal mycetomes.[187] These may be *Symbiotaphrina buchneri*[267] or *S. kochii*[265] or *Cryptococcus albidus* s.l. (unpublished observations). The symbionts are transmitted to the new generation via the egg after migration to the oviduct. When larvae hatch, they eat the shell and acquire the symbionts. These symbionts synthesize vitamins, amino acids, and sterols, essential nutrients for *L. serricorne.*[186,265,303] Sterilization of the eggs in 5% chloramine in 70% alcohol[265] produces aposymbiotic larvae and beetles which require vitamin supplements.[146,266]

The production of esterases by the symbionts, such as 1-napthyl acetate esterase, may detoxify a wide range of pesticides, mycotoxins and plant toxins[118,311,312] and thus impart tolerance to *L. serricorne.*[119,242,303,312]

Laboratory diet

Several recipes have been developed for complex artificial diets providing optimal levels of nutrients.[101,185,255,267,316] However, for raising *L. serricorne* for experiments, other than those on diet, a simple mixture of (20:1) wheat flour and brewer's yeast is sufficient. More important than diet is the need to avoid inbreeding in the laboratory culture by introducing feral beetles to the culture every ten generations or so. The feral beetles should first be quarantined for two generations to ensure fitness and an absence of mites (section 2.1.5).

2.1.4 REPRODUCTION

Pheromones are external chemical secretions that influence the behaviour and/or physiology of individuals of the same species. The sex-attractant pheromones evolved to bring the two sexes of a species together for reproduction. In general, pheromones produced by the female are attractants while male pheromones are aphrodisiacs. Female sex pheromones are active over longer distances than those of the male, and are biologically active in minute quantities [280] (see section 3.2).

Female *L. serricorne* release a pheromone which attracts and excites males. [65,66,103,346] Pheromone production begins 10 hours after adult emergence, increasing to a maximal level 4 days later and gradually declining from the sixth day after emergence until death.[103] The responsiveness of males to the main pheromone component increases from the first to the fourth week of life.[226] The pheromone-producing gland is in the females' second abdominal segment.[228]

Extracts of intact pheromone glands are attractive to adult males.[228] Chemical study of the sex pheromone of *L. serricorne* [94,95,99,100] isolated seven active components (Table 2.2).[98]

Behavioural bioassays[99] and electroantennogram responses[98,99,245] indicate that **(4*S*,6*S*,7*S*)-serricornin** elicits the strongest sex pheromone activity, and is the main component of the sex pheromone which evokes strong attraction and sex stimulation in the males. Serricorone and serricorole[97] also stimulate strong sexual activity, but show a weak attractiveness. Anhydroserricornin may be merely a precursor or intermediary form of serricornin,[223,224,227,246] as it displays a magnitude of activity of less than one thousandth that of serricornin and displays no synergistic or inhibitory effects.

The raceme of 4*S*,6*S*,7*S*-serricornin, 4*S*,6*S*,7*R*-serricornin is, however, a pheromone antagonist, suppressing the activity of 4*S*,6*S*,7*S*-serricornin.[223,226,250] The rapid synthesis of 4*S*,6*S*,7*S*-serricornin[284] has led to commercial production of lures used to monitor *L.. serricorne* (Chapter 3). Serricorole also acts as an ovipositional deterrent[177] after being deposited on a substrate by ovipositing females.[202,204]

Mating occurs within 3 days of adult emergence from the pupal cell. When exposed to virgin females, 25% of unmated males mate by 3 days after emergence (ae), 70% by 4 days ae and 96% by 7 days ae. Response time to the presence of virgin females decreases with age, from 7 minutes at 4 days to 4 minutes at 7 days.[226]

Following attraction to the female, males lower their head and upper body (prothorax), vibrate their antennae and rapidly walk around the female. Prior to copulation, the male touches the female with his antennae and palpi, grasps her elytra and inserts the aedeagus into the genital opening. The male descends and remains paired with the female in an end-to-end position for about 1 hour. During this stage copulation occurs and sperm transfer takes place. Males remain responsive to females after previous matings.[226] Multiple matings occur, with over 90% of females able to mate at least twice and over 90% of males able to mate at least six times.[102] Although multiple matings have little effect on female fecundity, almost 85% of progeny develop from sperm obtained at the second mating. Mating success may be due to female weight,[102] but any effect of female age is disputed.[218]

Table 2.2 Components of the sex pheromone of *L. serricorne*[98]

(4*S*,6*S*,7*S*)-7-hydroxy-4,6-dimethyl-3-nonanone (serricornin)
2,6-diethyl-3,4-dihydro-3,5-dimethyl-2H-pyran (anhydroserricornin)
4,6-dimethyl-3,7-nonanedione
4,6-dimethyl-3,7-nonanediol
7-hydroxy-4.6-dimethyl-4-nonen-3-one
(2*S*,3*R*)-2-ethyl-2,3-dihydro-3,5-dimethyl-6-(1-methyl-2-oxobutyl)-4H-pyran-4-one (serricorone)
(2*S*,3*R*)-2-ethyl-2,3-dihydro-6-2-hydroxy-1-methylbutyl)-3,5-dimethyl-4H-pyran-4-one (serricorole)

2.1.5 NATURAL ENEMIES

The most important parasites of *L. serricorne* are the wasp, *Anisopteromalus calandrae* (How.), and mite, *Moniezella angusta* (Banks).[23] *A. calandrae* develop from egg to adult in 16 days, the female wasps mate and then oviposit in beetle larvae and pupae. *M. angusta* kills and feeds on the larval and pupal stages. Females lay eggs loosely and promiscuously, and their population increase is rapid.[23] Other recorded parasites and predators are shown in Table 2.3. None of these enemies (Table 2.3), or others associated with stored products[54] have shown promise for *L. serricorne* control.

2.2 *EPHESTIA ELUTELLA* – THE TOBACCO MOTH

The tobacco moth, *Ephestia elutella* (Hübner), (Lepidoptera: Pyralidae), was first recorded as a pest of cured tobacco in Russia in 1915.[247] Also known as the cocoa or warehouse moth, *E. elutella* will feed on a wide range of stored products including tobacco.[111,173] The evolutionary origins of *E. elutella* are thought to be in the Northern hemisphere temperate zone, due to its readiness to enter diapause[259] and exclusion from the

Table 2.3 Natural enemies of *L. serricorne*[173]

Name	*Type*	*References(s)*
Acaropsis docta	Predatory mite	5
Acaropsis solers	Predatory mite	294
Anisopteromalus calandrae	Parasitic wasp	2, 18, 23, 25, 51, 110, 191, 230, 285, 319
Bacillus cereus	Bacteria	344
Bruchophagus sp.	Parasitic wasp	231
Cephalonomia gallicola	Parasitic wasp	130, 191, 230, 371
Chaetospila elegans	Parasitic wasp	23
Cheyletus eruditus	Predatory mite	23,281,297
Chortoglyphus gracilipes	Predatory mite	21
Israelius carthami	Parasitic wasp	290
Lariophagus distinguendus	Parasitic wasp	23,25,221
Moniezella angusta	Predatory mite	23
Norbanus sp.	Parasitic wasp	183,195
Nosema plodiae	Protozoa	363
Pediculoides ventricosus	Predatory mite	23, 69, 230, 251, 285, 326
Pyemotes tritici	Parasitic mite	60
Pyxinia sp.	Protozoa	279
Rhagidia sp.	Predatory mite	183,198
Seiulus sp.	Predatory mite	23,230
Tenebroides mauritanicus	Predatory beetle	19, 23, 68, 69, 230, 285, 319, 326, 343
Thaneroclerus buqueti	Predatory beetle	69, 182, 183, 195, 230, 249, 297
Tribolium castaneum	Predatory beetle	179, 211
Uloborus geniculatus	Predatory mite	220
Venturia canescens	Parasitic wasp	257

tropics due to intolerance of long exposures to high temperatures.[111] The tendency to diapause (dormancy) at lower temperatures or short photoperiods allows only one generation in cool temperate climates,[33,34] although up to three generations per year have been reported in warmer climates.[173] A recent review[16] was derived from this chapter and appears here with kind permission from Elsevier Science Ltd, UK.

2.2.1 MORPHOLOGY AND LIFE HISTORY

Adults

Adults (Plate 2.8) are small grey or brown-grey moths, measuring just under 1.0 cm from the head to the tips of the folded wings. Wings have a spread of about 1.6 cm.[173] Females weigh between 9 to 11 mg and males weigh between 7 to 9 mg, depending on food type.[358] When reared on the same substrate, male moths are always lighter in weight than females.[358] Identification of *E. elutella* requires a close examination of the genital structures,[106] although identification from external characters is possible.[30]

When using pheromone traps baited with ZETA (section 2.2.4) one is likely to capture *Plodia interpunctella* (Plate 3.10) or other *Ephestia* species such as *E. kuhniella*. During the day, the moths rest against the ceiling or walls, and become active at dusk.[306,360] Adult moths move towards heat and react to temperature differences of 3°C: this attraction overrides the response to move away from light.[360] In the warehouse adults live up to 3 weeks, with males rather shorter-lived than females.[291,361] Copulation takes place 1 or 2 days after emergence, and may occur several times during the life of the moth.[306] Females lay more than 100 eggs,[111,173] mostly at dusk or during the night in the first 4 or 5 days of life.[36,306,360]

Eggs

Eggs (Plate 2.9) are laid singly or in groups on or near the food source.[173] They are sandy white when laid, turning darker as the embryo develops. Each egg is ellipsoid, weighs about 3 μg and measures 0.5 mm long and 0.35 mm wide. The egg shell is tough, finely ridged[13] and only loosely attached to the substrate. The time taken for eggs to hatch varies with temperature, but at 25°C the larvae emerge within 6–7 days.[111]

Larvae

Larvae (Plate 2.10) reared on tobacco vary in colour between light yellow, brown, pink or off-white.[3,173] Larvae generally develop through six instars,

Plate 2.8 *E. elutella adult*, dead and displayed to show the wing pattern. At rest the moth folds the wings over the abdomen.

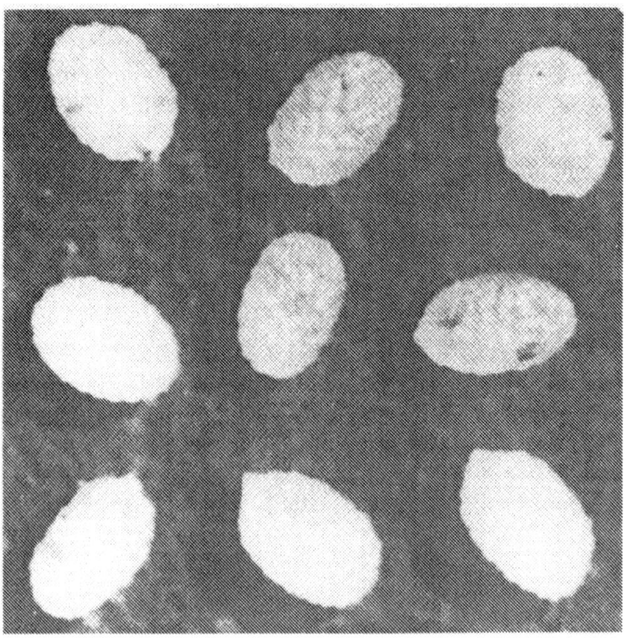

Plate 2.9 *E. elutella* eggs.

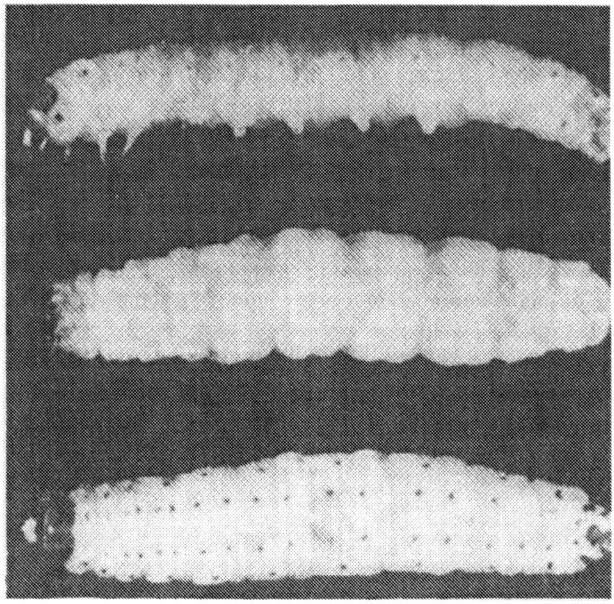

Plate 2.10 *E. elutella* larvae.

and are 13 mm long when fully grown.[173] After hatching, larvae burrow into the tobacco, begin feeding and remain unseen for 2 to 3 months. Only the larvae feed on tobacco, attacking primarily flue-cured and oriental tobacco. The moth larvae are heavier feeders than larvae of *L. serricorne*. In a severe infestation entire leaves except for the midrib and larger veins may be devoured.[173] Full-grown larvae enter a wandering phase, spinning silken threads behind them as they wander and emerge from the tobacco. These silk threads, which may form sheets (see section 3.1), are indicative of infestation.[291,306] In a warehouse the larvae often move via the walls to the ceiling or a crevice where they spin a silk cocoon and may diapause for 4 to 9 months.[38,39,173,291,359] Larvae may pupate and emerge to produce a second generation before the onset of winter, depending on the climate[291,306]

Diapause in insects is "dormancy", closely linked to the environment which acts as a synchronising mechanism to ensure that breeding occurs at the optimal time of the yearly cycle.[41] The presence of a small number of *E. elutella*, pupating early and forming a second generation may be regarded as an adaptation to a stable, yet uncertain environment. With most of the population protected in diapause, a few are still active and able to exploit the possibility of a warm winter.[41] Diapausing larvae are not completely inactive, but metabolize slowly and produce solutions of nitrogenous waste continuously.[359]

The larvae pass through three stages during diapause: a true diapause, where high temperatures do not induce pupation; a quiescent period, in which pupation is delayed by low temperatures and initiated by high temperatures; and a prepupal stage, characterized by the release of a pupation hormone, more abundant excretion, final emptying of the gut and the final moult and pupation.[359] The most efficient completion of diapause in *E. elutella* is dependent on periods of low temperature, followed by a sustained period at high temperature. Long photoperiods improve survival by helping to synchronize pupation. Long photoperiods also assist the completion of the cold-dependent phase of diapause if the exposure to a low temperature is too short and the temperature following this cold exposure is high enough.[39] If the temperature following the cold exposure is too low, even in long photoperiods, emergence may become asynchronous and mortality increases.[39,41]

Pupae

Sixth instar larvae eventually pupate. Pupal colour changes from light to dark brown, with the pupae (Plate 2.11) becoming darker as they get older.[173] The duration of the pupal stage is governed by temperature, and varies from 45 days at 15°C to 10 days at 30°C.[31] Pupae are always found

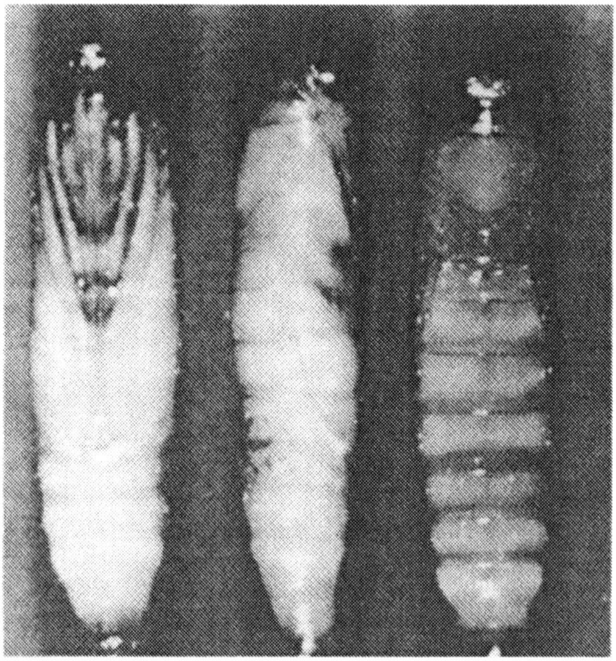

Plate 2.11 *E. elutella* pupae.

on external surfaces since the emerging adult moth is fragile and, unlike *L. serricorne*, cannot push its way out of confinement.[173]

2.2.2 SURVIVAL AND DEVELOPMENT

During diapause the moth is tolerant of extremes in temperature, humidity and even some insecticides. *E. elutella* larvae have been reported to survive the winter in unheated buildings in Canada at around –1°C.[173] At 70% RH, the temperature range over which development can be completed is from about 15°C to 30°C[31] (see Chapter 5 and Fig. 2.3). The rate of development varies with temperature: the egg to adult period would be about 50 days for a temperate summer.

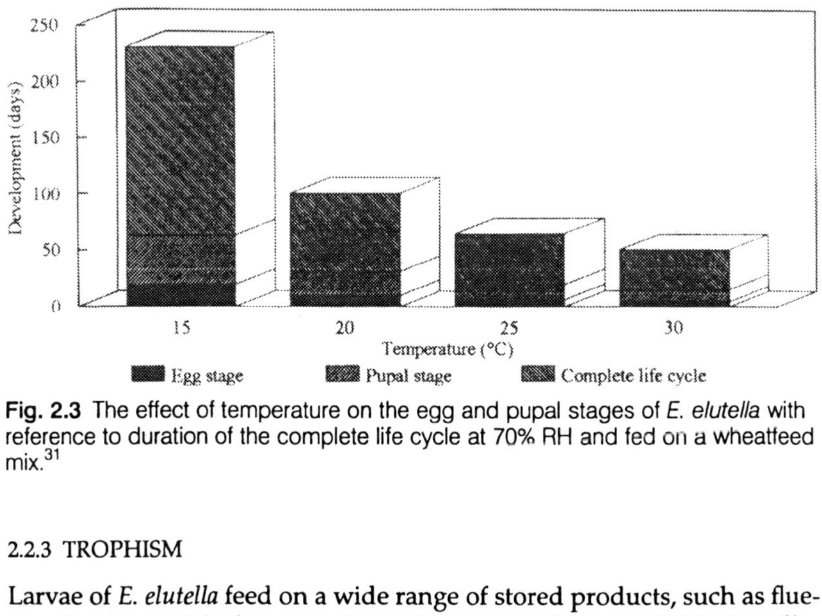

Fig. 2.3 The effect of temperature on the egg and pupal stages of *E. elutella* with reference to duration of the complete life cycle at 70% RH and fed on a wheatfeed mix.[31]

2.2.3 TROPHISM

Larvae of *E. elutella* feed on a wide range of stored products, such as flue-cured and oriental tobacco, grain products, cocoa beans, dried fruit, coffee and cayenne pepper.[146,173]

The moth is not recorded infesting air-cured, fire-cured and cigar types of tobacco and has rarely been recorded infesting manufactured tobacco products. It, like *L. serricorne*, prefers tobacco with a high sugar and low nicotine content.[173]

Food requirements

E. elutella differs from *L. serricorne* in that it has no recorded intracellular symbionts, and requires a source of B-group vitamins and other nutrients. It is usually raised in the laboratory on flour[25,146] or more exotic diets.[147,148,316,322,330] When rearing *E. elutella* care must be taken to ensure an adequate food supply[32] for larvae and space for adult mating.[50]

2.2.4 REPRODUCTION

Both male and female *E. elutella* produce **pheromones** associated with mate attraction and mating behaviour. The females produce (Z,E) 9,12-tetradecadien-1 ol acetate (ZETA) from a gland between the eighth and ninth abdominal segments,[207] which acts as a long-range attractant and stimulates the mating behaviour of males.[207,334] ZETA is found in several other species of stored-product moths, and due to the cross-attraction between species,[58,271,334] it is likely that a more specific response is then elicited by a short-range male pheromone as well as specific courtship behaviours.[206,334] The males produce a pheromone from specialized wing gland scales on the fore wings,[206] not from glands on the eighth abdominal segment.[115,206] This male pheromone is (E)-phytol (3,7,11,15-tetramethyl-2-hexadecenyl alcohol), the most important component, and a series of saturated *gamma*-lactones.[272] This elicits a stereotyped courtship response from females of the same species.[206,272]

Adult *E. elutella* have characteristic sexual calling behaviour[207] but no mating noises.[368,369] Females flex the abdomen dorsally, evert the ovipositor, and beat their antennae 80 beats per minute.[207] Courtship proceeds as the male *E. elutella* exposes the scent glands, walks toward the calling female, fans his wings, and manoeuvres into a head-to-head orientation with her so that the female's antennae are close to the male's scent glands. The female lowers her abdomen out of the calling position, withdraws her ovipositor and pheromone gland, flexes her abdomen ventrally, up and down several times prior to the male's attempt to engage the female's terminalia.[206] The male inserts his aedeagus into the genital opening of the female, and the pair move into an end-to-end position where sperm transfer takes place. Courtship success has been correlated to male size, with females actively discriminating against males of small size. By mating with larger males, females are both more likely to produce larger offspring and increase their fecundity.[271]

The pre-oviposition period following pairing lasts, for most females, over 24 hours. Most of the eggs are laid in the first 4 days following copulation. *E. elutella* prefer rough to smooth surfaces for oviposition, and most eggs are laid in the evening or at night. Ovipositing females show preference for surface type, and show strong chemotaxis towards cacao beans, and less towards grain.[360]

2.2.5 NATURAL ENEMIES

The most important natural enemies of *E. elutella* (Table 2.4) are *Bracon hebetor* (Say) and *Seiulus* sp. and *Cheyletus* sp. mites.[23] *B. hebetor* is a wasp that stings and paralyses the larger *Ephestia* larvae and then lays eggs on them. The eggs hatch and the larvae feed externally until they become mature. *Seiulus* sp. mites are abundant and universal, preying on the *E. elutella* eggs[23] and adults.[230] Mottled mites (*Cheyletus* sp.) are predatory on the smaller *E. elutella* larvae and may be of considerable importance in their natural control.[23] Adult mites attach themselves to the larvae, which become paralysed by secretions from the large cheliceral glands, and the mites then suck the body fluids from the larvae. Pink mites (also *Cheyletus* sp.) are predatory on *Seiulus* sp. mites and the larval stages of *E. elutella*.[23]

2.3 OTHER INSECTS REPORTED INFESTING TOBACCO

Many insects have been reported on tobacco, usually as predators or

Table 2.4 Natural enemies of *E. elutella*[293]

Name	Type	Reference(s)
Angitia chrysosticta	Parasitic wasp	360
Bacillus thuringiensis	Bacteria	50,64
Bracon hebetor	Parasitic wasp	23,236,360
Cheyletus sp.	Predatory mite	23
Holepyris hawaiiensis	Parasitic wasp	257,292
Idecthis canescens	Parasitic wasp	23,230
Laelops sp.	Predatory mite	23,230
Mesostenus gracilis	Parasitic wasp	23,230
Nemeritis canescens	Parasitic wasp	360
Orius insidiosus	Predatory bug	23,230
Persisierola gallicola	Parasitic wasp	314
Piezostethus flavipes	Predatory bug	57,292
Salticus scenicus	Spider	23,230
Seiulus sp.	Predatory mite	23,230
Theridion tepidariorum	Spider	23,230
Tricogramma sp.	Parasitic wasp	50

parasites of *L. serricorne* or *E. elutella* (Tables 2.3 and 2.4). Most reports of such infestations are prior to the widespread use of phosphine (Chapter 8) when true infestations were more severe. Indeed, today one may find a whole variety of predators, parasites, scavengers, and accidental pests on tobacco that has not yet been subject to a phosphine fumigation, such as carry-over tobacco from farmers (section 4.3.1, and Plate 4.3).

There are limited reports, usually from the tropics, of other insects actually able to survive on a diet of dried tobacco. Again, these predate the 1950s and even then were not commonly reported. No reports of infestation due to these insects have appeared since 1970. Other dried tobacco feeders are limited to the *Tycorynus tabaci* (Guerin–Meneville),[297] the larger tobacco beetle (greater cigarette beetle; tropical cigarette beetle; synonymous with *Catorama tabaci*). A native of Cuba, this was imported to the USA and France.[297]

Another insect that possibly feeds on dried tobacco is *Tlascala finitella* (Walker).[339] Other insects able to feed on tobacco seeds include: *Catorama impressifrans* (Fall); *Attegenus piceus* (Olivier); *Mezium americanum* (La Porte).[286,297]

Maintain vigilance in case other insects become a problem such as booklice (Psocids) in Zimbabwe.

Monitoring 3

When we know the distribution and abundance of *L. serricorne* and *E. elutella* throughout the various steps of the tobacco process (Fig.1.1) we may take appropriate preventive and control actions (Table 1.1). **Accurate monitoring data drive our control actions**. The traps used are not a means of control themselves as was once hoped[173,287,297] or indeed as sometimes suggested,[66,222,252] although, control via mating disruption by pheromones is feasible with other insects.[184]

3.1 VISUAL INSPECTION OF TOBACCO

When a shipment of tobacco is received, inspect it for the presence of infestation before introduction into the storage facility or factory. Inspect between 3 and 10% of cases with good lighting and have a hand lens available.

Heavy infestations of *L. serricorne* are easily detected as the tobacco has the characteristic "shotgun" peppering of larval burrows (Plate 3.1). Light infestations are more difficult to detect but the presence of beetle excreta on individual leaves may be seen with the naked eye and confirmed with a hand lens. Equally, larvae, pupae, or adult beetles may be found by carefully raising the individual, surface leaves.

E. elutella, in their final stage of larval development, are approximately 13 mm long and leave the tobacco to pupate; therefore, they are easily seen in infested tobacco along with the heavy damage they cause (Plate 3.2). The presence of silk threads on the outside of bales, entrapping pellets of excrement, is also characteristic of *E. elutella* infestation (Plate 3.3). Hold the tobacco bale on its edge and blow lightly to demonstrate the presence of these silk threads.

3.2 PHEROMONE TRAPS

3.2.1 *L. SERRICORNE* MONITORING

The discovery of the female's sex-attractant pheromone and its subsequent synthesis (section 2.1.4) led to the development of

Plate 3.1 Typical shotgun appearance of tobacco infested by *L. serricorne*. Shotgun holes are holes burrowed by the larvae as it eats or by adults as they emerge from pupal cases.

Plate 3.2 Typical appearance of tobacco infested by *E. elutella* where the larvae have eaten the leaf to the midrib.

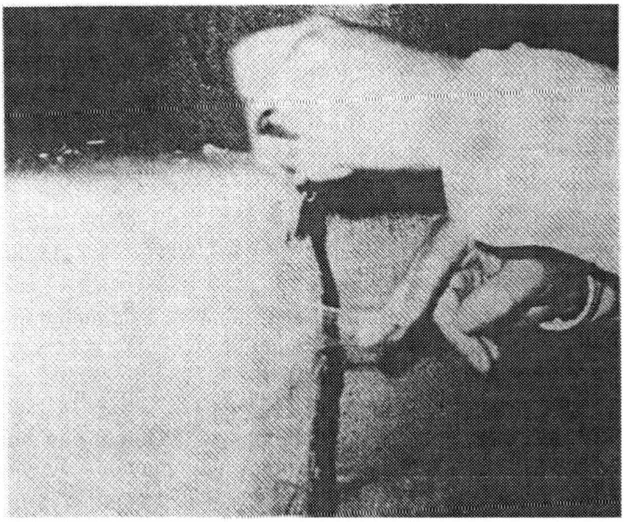

Plate 3.3 Silk threads and silk sheets on the outside of oriental tobacco bales, a sign of heavy infestation by *E. elutella*.

pheromone-baited traps. The essential element of the trap is a lure permitting controlled release of sex pheromones. The lure is mounted on a sticky board[222] that catches the beetles for counting.[22,133]

Three commercial sources of pheromone lures are available. They use 4*S*,6*S*,7*S*-serricornin as the main component and have different proportions of the minor pheromones (Table 2.2, section 2,1,4). The three lures are the New Serrico®[96] which is also used with the Sanitrap®trap, the Lasiotrap® lure[165,228] and the Trécé lure. Lasiotrap lures are changed, as recommended by their manufacturers, every 4 weeks, New Serrico lures every 8 weeks, and Trécé lures every 6 weeks: the active life of the lures, respectively. These traps catch greater than 90% males.[222]

The Lasiotrap (Plate 3.4) trap has three exposed adhesive surfaces (each 80 × 150 mm) and the pheromone lure in a closed, plastic tube. The New Serrico trap (Plate 3.5) is made up into a slim box (180 × 80 × 7 mm) to protect and enclose the adhesive surfaces (Plate 3.6); the beetles enter the trap via the open sides (Plate 3.7). In addition, the New Serrico trap is supplied with a food lure. The "New" was added when the high purity 4*S*,6*S*,7*S*-serricornin replaced the earlier formulation that contained the repellent raceme (section 2.1.4). The lure is impregnated cotton within a micro-perforated polyethylene film. The Sanitrap trap (200 × 70 × 14 mm) is similar in design to the New Serrico trap, but has closed sides with five (3 mm × 25 mm) slits in the sides to allow entry of insects, yet protect the adhesive surface from dust. If required, traps and lures may be purchased separately for assembly in any favoured combination. The Trécé Flit-Trak CB3® trap uses a pitfall trap and a rubber septa-type lure.

Pheromone trap advantages are; ease-of-use, increased sensitivity, small size, no electricity supply required, maintenance free, specific for *L. serricorne*,[280] mobile, and low cost of $2–5 each. A disadvantage is that traps with a sticky surface must be replaced when or if it becomes coated with dust. Pheromone traps have been used to judge the effectiveness of control actions such as aerosol application of insecticides in tobacco storage warehouses.[222]

For monitoring *L. serricorne* populations, the pheromone traps are spaced about 20 m apart or one per 3,000 m^3. Traps are placed on walls, machines, or structural supports, preferably with a ledge beneath, to allow the beetle to alight and crawl in. Because of compact size and ease of placement, the density of pheromone traps can easily be increased to indicate the precise location of infestation sources.[132,165] In the factory and warehouse environment, such precise location allows isolation and/or treatment of small areas and removes the need to disinfest a whole store or processing area.

Pheromone traps have also been used to demonstrate the need to modify working practices. Examples of these modifications are detailed in section 4.2. As an example: in a cut tobacco store the level of trapped

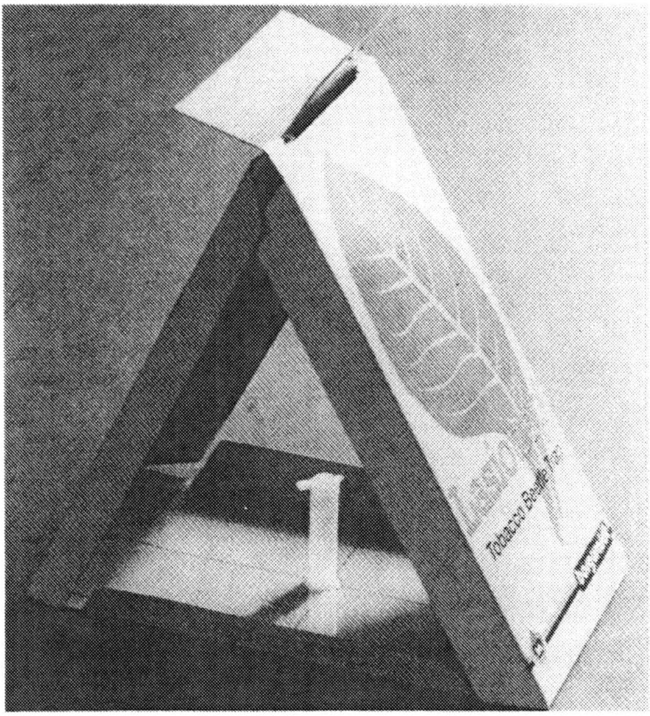

Plate 3.4 The Lasiotrap for *L. serricorne*.

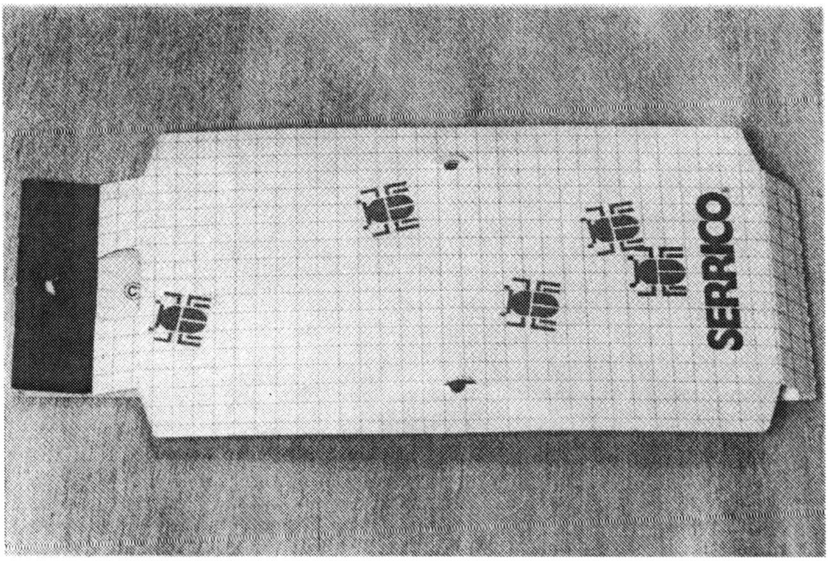

Plate 3.5 The New Serrico trap for *L. serricorne* closed and ready for use.

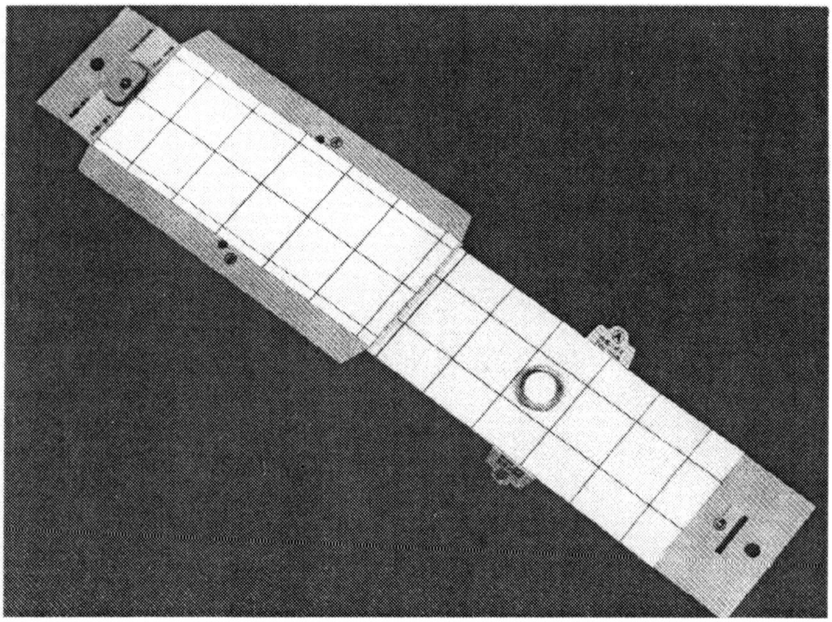

Plate 3.6 The New Serrico trap for *L. serricorne* open to show the glue surface, grid, and lure.

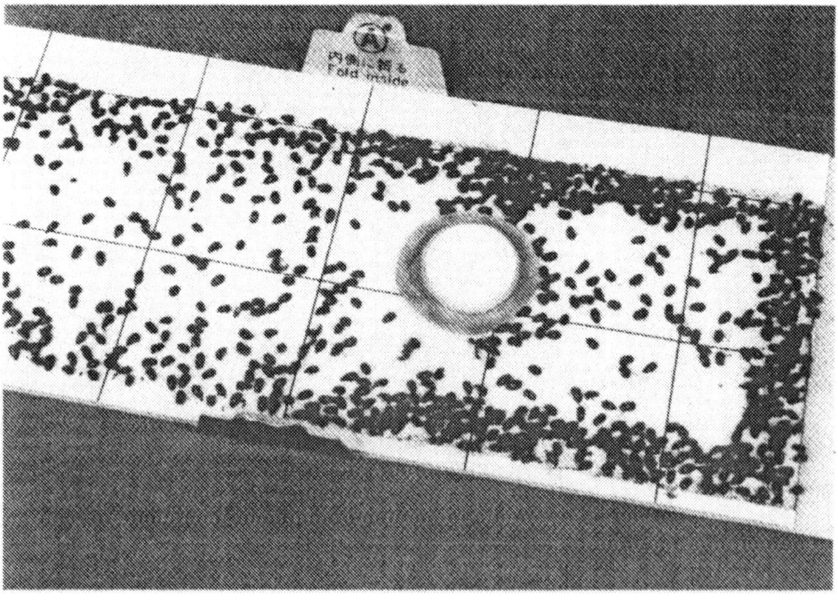

Plate 3.7 The New Serrico trap for *L. serricorne* showing numerous beetles caught on the glue surface.

beetles was five times higher in the 15 hours when the doors were open than in a 44-hour period when the store doors were closed and led to the installation of automatic doors.[93] Using pheromone traps to direct and drive the factory sanitation programme is covered in detail in section 4.2.2. Traps are usually protected (Plate 3.8).

3.2.2 *E. ELUTELLA* MONITORING

The discovery and synthesis of the female-produced sex pheromone, ZETA, (section 2.2.4) led to the production and use of numerous traps

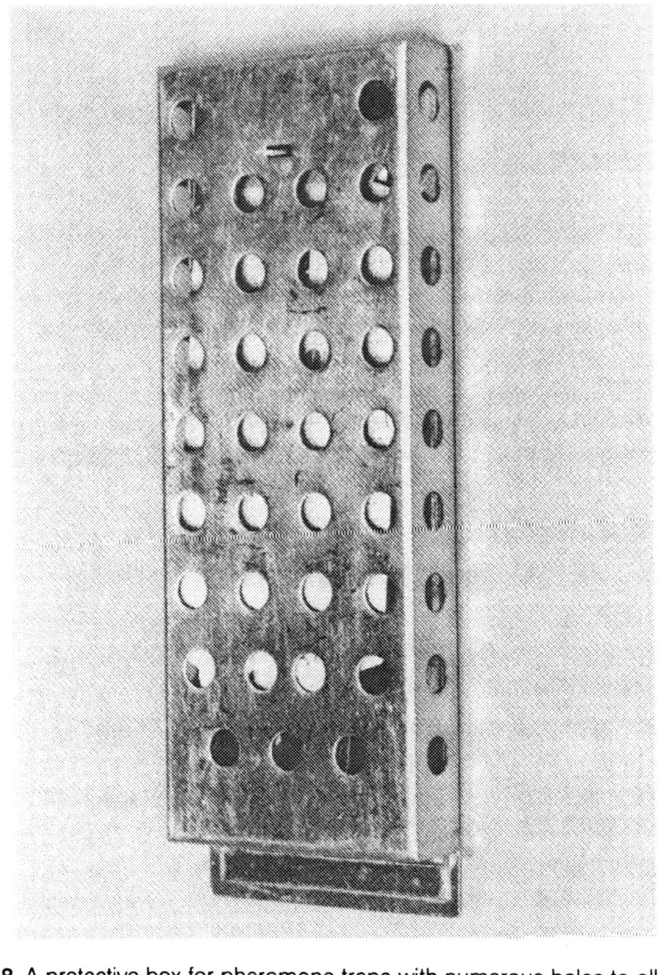

Plate 3.8 A protective box for pheromone traps with numerous holes to allow pheromone exit and beetle entry.

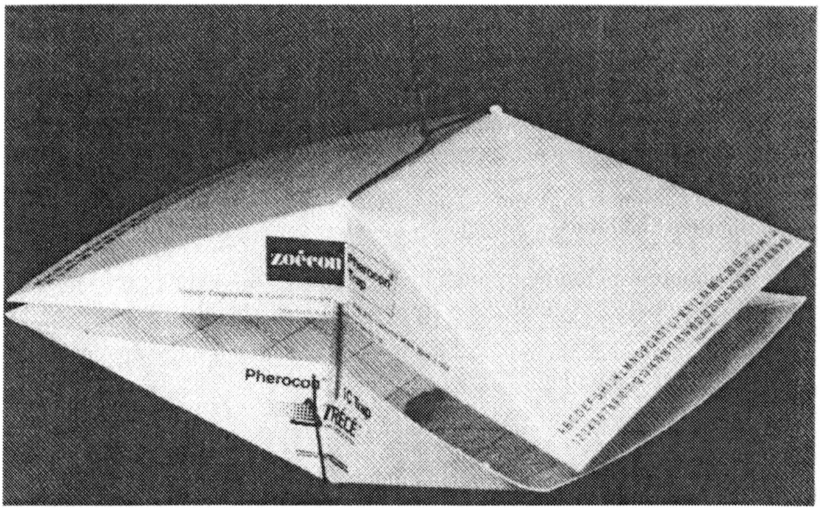

Plate 3.9 The Pherocon trap for *E. elutella*.

(Plate 3.9) for *Ephestia* species and the related Indian mealmoth (*Plodia interpunctella*) (Plate 3.10).[63,155] The various trap designs[155] have in common a lure and capture device. The lures may be plastic tubes, rubber septa or impregnated cotton. The capture device is usually a glue board or a one-way tunnel leading to a holding container. The design allows for the moths to alight from flight. Traps are placed, usually at 3 m from the ground to avoid head-contact and positioned 20 m apart or one per 3,000 m^3. They are used primarily in unprocessed tobacco storage or transit areas.

3.2.3 INSTRUCTIONS FOR USE OF PHEROMONE TRAPS

Basic instructions for using pheromone traps would include:

Glue boards are stored in a cool place to avoid glue running and are changed when covered with dust or covered with too many insects, making the counting ineffectual and difficult.

Lures are stored in a refrigerator between 6–10°C (not frozen) , removed from their pouch (foil wrap) just prior to use, and placed gently on the glue surface to avoid covering the slow-release opening or pores. Do not remove or touch the polyethylene cover.

Traps (glue board and lure) are to be used within 1 year of purchase, changed according to manufacturer's instruction, used when ambient temperatures are above 15°C in warehouses and all year-round in factories, placed one per 3,000 m^3 or 20 m apart, identified by a unique trap number, and counted on the same day each week.

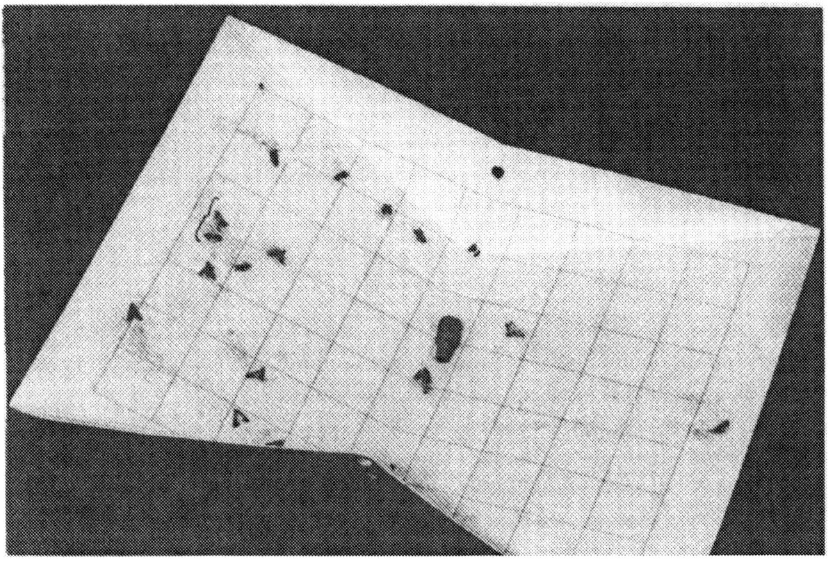

Plate 3.10 The Pherocon trap for *E. elutella* showing the lower glued surface, rubber septa, and captured moths. Several moths, for example the two in parenthesis, are clearly bicoloured and therefore *Plodia interpunctella.*

Counted means: count the insects caught the previous week; write the total caught on the trap and monitoring report; report the trap count to the responsible personnel to take action.

Inspect traps for the quality of the adhesive surfaces before use. Indeed, request QA data for glue boards and lures from the manufacturer. Records of trap counts including the date and location are under the co-ordinated control of one person on each site and retained to provide a record of the effectiveness of the control practices. Review these records regularly with the local management teams, and report any outbreak of infestation to the senior site manager. Establishing a pheromone trap monitoring programme is covered in section 4.2.4.

3.3 LIGHT TRAPS

The attraction of insects to light has been known since the times of early humans. The phenomenon was certainly recorded by 400 AD, but not until 1874 was the use of light traps formally reveiwed.[152] Photoreactive responses of *L. serricorne* were described in detail in the early part of this century.

L. serricorne shows a negative response to intense light, but is attracted to windows in the dim light of dusk.[297,299] The attraction of *L. serricorne* increases as the light source changes from the red end of the spectrum towards blue.[297] Unlike some insect species, *L. serricorne* preferred (78% versus 22%) UV light to green.[321] Light traps were used before the advent of phosphine fumigations when weekly captures were so high that *L. serricorne* were measured volumetrically.[287] Fortunately, infestations tend to be less severe today and require a more sensitive monitoring tool, the pheromone trap.

3.3.1 SUCTION-LIGHT TRAPS

The combination of suction traps for *L. serricorne* operating near light traps had begun by 1920,[297] though the combined suction-light trap was not fully introduced to the tobacco industry until 1934.[287]

A full description and diagrammatic representation of the standard suction-light trap used in the tobacco industry, until the advent of pheromone traps, is published.[287]

Suction-light traps are easily constructed in metal workshops and with non-specialist materials. Incandescent light bulbs are effective in attracting *L. serricorne* and *E. elutella* and a power between 40 and 60 watts has been proposed as optimum.[287] Less than 40 watts does not attract and more than 60 watts repels. To increase trap efficiency of this standard suction-light trap the incandescent bulb was replaced with a 32-watt circular fluorescent black light mounted in the front part of the flange.[320,350] An electrically driven fan (about 40 W) sucks the attracted insects into the collecting jar.

3.3.2 ELECTRIC GRID-LIGHT TRAPS

Light traps in which insects are attracted and killed when they make contact with an electrified wire grid have also been used;[138] however, the standard light trap is simpler and easier to build and maintain.

3.3.3 LIGHT AND ADHESIVE COMBINATION TRAPS

These could be considered the most simple of light trapping devices and are often only an electric bulb with an adhesive collection/trapping tray mounted a short distance beneath the bulb.[154]

3.3.4 USE OF LIGHT TRAPS

Light traps are usually spaced every 3,000 m^3 [287] to 8,000 m^3.[350] Because light traps are large, require an electricity supply, and need to be positioned to avoid collisions with moving machinery, they are, generally, suspended from the ceilings using a pulley system.

3.4 COMPARISONS BETWEEN TRAPS

Most of the studies to compare different traps are out-of-date, and it is recommended to make trials before choosing a particular trap. **Pheromone traps are now the industry standard** and most perform equally well. Choice tends to be based on cost, availability, ease-of-use, and the manufacturer's support.
Five examples are given below of some of the trap comparison studies.

L. serricorne

1. Light traps (30 W black light), pheromone traps and blank sticky traps without the pheromone lure were located at four positions, and exchanged weekly to ensure that trap location did not influence the results. The pheromone capsule was renewed monthly and *L. serricorne* counts made weekly. For every beetle caught in the light trap, between 10 and 15 were caught in the pheromone traps.[165]
2. A light trap (200 W incandescent) and sticky surface was compared to an "old" Serrico trap (i.e. + *4S,6S,7R*), a Lasiotrap, and a Serrico trap without the lure. All traps were positioned 1.5 m high and 16 m and 20 m apart and positions rotated weekly. The light trap was reported as being most efficient.[154] The difference between the above studies may be due to the different bulbs used. However, the unusual placing of a light trap 1.5 m off the ground would not provide a practical arrangement in the working environment. Equally, the New Serrico trap is now greatly improved. Finally, the finding of attraction to a 200-W light source contradicts earlier work that states insects are repelled above 60 W (section 3.3.1).
3. A comparison of black light traps, a Serrico pheromone trap and a sticky trap[132] showed the "old" Serrico trap to be superior to the other traps. However, unlike the above studies,[154,165] insufficient details are given to determine the circumstances under which the traps were tested.

E. elutella

4. In comparisons of various moth pheromone traps to a light trap (200 W) and sticky trap, it was found that the pheromone traps captured up to seven times more *E. elutella* than the light trap.[155] In addition, the design of the pheromone traps was critical in determining the capture rate. Basically, if glue boards are used with the lure, these must be changed if covered with moths, whereas, no-return plastic containers used with the lure are emptied and re-used.
5. During a 2-year survey of *E. elutella* in warehouses in Piraeus, Greece, the pheromone traps caught 6.4 times more moths than the glue board alone, without the ZETA lure.[63]

3.5 ACTION THRESHOLDS

A desirable goal for all tobacco-related operations is zero insects caught in pheromone traps and none seen during visual and/or sanitation inspections (section 4.3). Achieving this goal is clearly easier in the colder temperate regions, where *L. serricorne* may have only three or less generations a year, than in the tropics, where year-round breeding allows six or more generations per year. Conversely, the temperate zone is the natural habitat for *E. elutella*, thus tobacco in these regions is more at risk.

We do not propose global eradication of *L. serricorne* or *E. elutella* as this is clearly unobtainable. One must strive to maintain population levels as low as is realistic for a given operation. A zero count in a factory does not mean eradication.

Zero tolerance is our goal, and it **is** achievable. If it is already achieved, we must maintain preventive measures. But how do we use the pheromone trap data to direct sanitation and other control actions, to reduce infestation to zero?

Action thresholds are based on trap monitoring data that must be collected for at least 6 weeks. More data means the calculated action thresholds will be more realistic. Calculation is best based on accepted statistical process control parameters.[125] A more simplistic approach would be to use the mean, thus, if pheromone trap counts for the prior 6 weeks were: 5, 7, 3, 9, 4, 8, the mean = 6. The mean (6) would be the subsequent action level, and provoke more intensive sanitation leading to a lower mean, or action level, for the next period and so on. Once one reaches near zero levels the actual trap catch is used until zero is achieved.

In areas having an indigenous population of *L. serricorne*, the setting of absolute thresholds upon which to take action may be problematic. Much depends upon local circumstances. For example, finding any beetles in a regularly fumigated warehouse may indicate an infestation problem. However, near a *L. serricorne*-infested cocoa warehouse, regular finding of beetles in traps may well occur until screens are installed. The capture levels, therefore, should be used in conjunction with the local circumstances, historical trapping data, seasonal variations, common sense, and visual inspections to determine when *L. serricorne* or *E. elutella* is a real threat to the tobacco and product integrity.

Requirements or standards must be set for each of your operations. Action thresholds are such requirements and allow conformance measurement. This type of approach is routine in quality management.

Sanitation 4

As tobacco journeys from field to factory to consumer (Fig. 1.1), it must be accompanied by a parallel regime of sanitation and pest control. The three principles, prevention, control and elimination, are the cornerstone of an integrated pest management programme which every pest controller or sanitarian will pursue. This chapter addresses these aspects, concentrating on prevention through a policy of sanitation.

Insect pests are a primary concern of this book; however, it is important to mention mould spoilage. There are parallels to be drawn and lessons to be learned from mould prevention. Like insect pests, certain species of mould have established a niche through their ability to germinate on a relatively dry substrate such as tobacco. Sanitation, along with moisture control, will prevent the growth and spread of spoilage organisms on tobacco.

4.1 PHILOSOPHY

The tobacco industry has high product standards and the development of a sanitation standard is an obtainable goal. This standard may exist as a separate entity but it is better placed within the overall quality system. Such standards are recorded as part of the quality system within quality management and may be required by health and safety-related law. These standards may be part of quality assurance (QA) accreditation via programmes such as ISO 9000[178] or its national counterparts where the sanitation standard could be structured as shown in Fig. 4.1. Such a standard would be known as the company or **Factory Sanitation Programme (FSP)**; also see section 9.8.

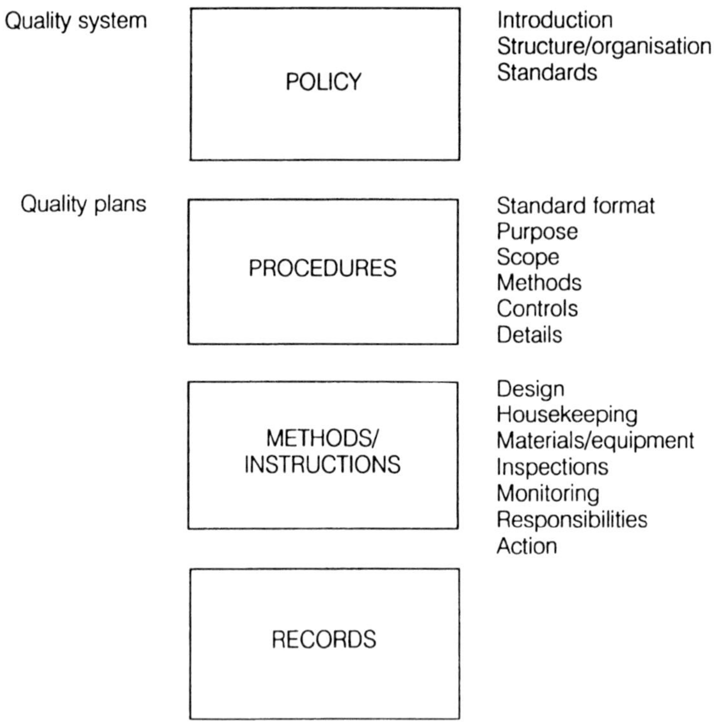

Fig. 4.1 Suggested structure for a company sanitation standard.

The sanitation policy, as with safety or quality systems, details the organisational structure, responsibilities, procedures, processes (or methods) and resourcing for implementing the sanitation objectives. A simpler quality plan, which is more task-orientated, could be developed under an overall, pre-existing quality system such as ISO 9000 or its counterparts. The plan would detail the sanitation objectives and acceptance standards with specific procedures for their enactment. A programme of inspection or audit to required and recorded standards is essential, as is a third party or independent examination.

A named coordinator ensures that correct, prompt action is taken.

Such systems need not be limited to prevention but may be extended to all activities to include control and elimination. In any system, a corrective loop is established with regular reviews and updating of standards and systems (Fig. 4.2).

The *requirements* and *benefits* are the same whether it is safety, quality, or sanitation:

Sanitation management : *requirements*
- Policy – commitment from management.
- Responsibilities – clear allocation with authority.

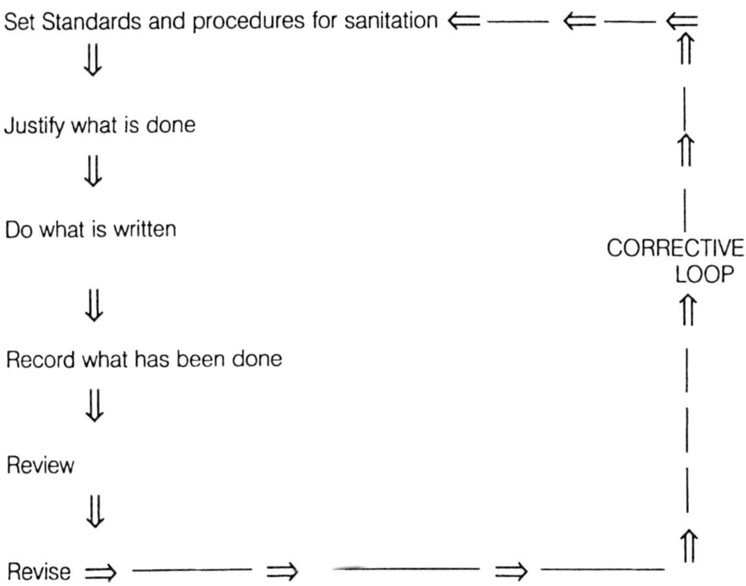

Fig. 4.2 The "corrective loop" system applied to sanitation.

- Pre-planning.
- Documented procedures and practises.
- Assessment of suppliers/contractors (where applicable).
- Internal inspection/external audit.
- Reviews
- Training/communication/feedback

Sanitation management : *benefits*
- Improved product quality: less infestation, spotting, off-taste problems.
- Reduced product contamination.
- Minimal costs and improved profitability.
- Better employee morale and motivation.
- Safer, improved workplace.
- Increased productivity: clean machines are more efficient.
- Minimal use of chemicals for pest control.

The parallels between sanitation, safety and quality management are primarily in the common culture established: that is one of preventive management through planning, set objectives, audits and reviews.

The adoption of a written sanitation policy or system is recommended and how to develop this is outlined in section 4.2.4.

4.2 PREVENTION

Factors affecting the growth of established pest infestations include the availability of suitable food, undisturbed shelter, and suitable environmental conditions. The processing of tobacco generates tobacco dust which drifts within the the building and settles on ledges, pipework, beams, machinery (outside and inside), ducts and floor voids. This drift of dust to hard-to-find areas is exacerbated by the use of compressed air as a cleaning tool (this is **not** sanitation). This dust may provide sanctuary and food for *L. serricorne*. However, conditions suitable for pests and their development will not normally arise if the principles of preventive pest control are observed.

The objectives of preventive pest control are to make the premises inaccessible and unattractive to pests and to stop pests getting at the raw materials and finished products, while at the same time protecting the buildings and providing comfort and safety to the workforce. Activities which may help reach these objectives include cleaning, building maintenance, and good storage practise.

The simple principle of good housekeeping is essential in infestation control. It should be regular, thorough and efficient. Effective cleaning requires a schedule which specifies the frequency, methods, equipment and responsibilities of key personnel. All of this should be set out in the **Factory Sanitation Programme** (see Fig. 4.1). Most ideas behind building maintenance and storage practice are common sense, but require proper co-ordination with cleaning. Cleaning is expensive: it consumes a lot of time, energy, manpower and materials. It must be made cost effective by optimizing the cleaner's time, selecting and using the correct materials for the job and effective training of the staff. **Removing a long-term *build-up* of tobacco residue takes much more time, energy, and materials than does a simple clean-up programme on a regular basis.** Identify the true cost of sanitation efforts and this data will be used to justify investments to negate inefficient sanitation.

Regular inspection of all raw materials, storages, manufacturing areas and finished stock is essential in detecting pest damage and in ascertaining the effectiveness of the preventive sanitation programme. Planned inspections are a feature of every preventive programme and can be structured in a variety of ways, such as departmental checks, location inspections by company sanitarians, or full scale external audits. Where infestation is suspected or known, inspection with monitoring techniques such as pheromone traps (Chapter 3) may help to assess the preventive measures and to assist in coordinating control measures. In other words, pheromone traps will show a problem exists, locate the exact problem area, and indicate the effectiveness of cleaning.

4.2.1 CLEANING

The work of the cleaning staff must be facilitated. In particular, improve access because areas difficult to clean will probably not be cleaned properly, if at all. Examples of such areas are: gaps between storage pallets and walls, the spaces underneath and behind machinery and fixtures, and the "dead" spaces in machines. Dead spaces in machines must be cleaned. Raise machines sufficiently clear of the floor (at least 10 cm) to assist both regular and deep cleaning.

Remove redundant items of equipment, worn-out machinery and unwanted fixtures as they harbour dirt and tobacco residues, form potential breeding sites for tobacco pests, make cleaning operations more difficult, and take up valuable space. Immediately deep clean machinery which is *temporarily* unused to remove all tobacco residues so it is ready for later use on site or elsewhere. One may cover such unused machinery with a sheet and monitor this space with a pheromone trap.

Pay regular attention to drains, gullies and sinks, for they can be an important source of mould, flies, cockroaches, other pests and rodents. Clean and inspect frequently and remedy any defects found without delay. Remove waste materials and tobacco which accumulate in bins, etc. every day. Clean bins regularly and store no tobacco on the production floors.

Brief all cleaning personnel as to the nature of tobacco pests and encourage them to report their finding of anything suspicious indicating the presence of insects. The objective is to *underline* the importance of their work in contributing to sanitation and pest control and to gain their help in the early detection of infestations. **Industrial vacuum cleaners** (Plate 4.1) with numerous suitable attachments (Plate 4.2), will be necessary to clean efficiently. *Air blowing has no place in a sanitation programme: this exacerbates the situation by causing dust movement to inaccessible areas.* Suitable vacuums have a 3-kW motor capable of moving 7 m^3 of air per minute. Attachments needed will be quickly identified by the cleaning personnel, to enable them to meet the cleaning specification. Each department must prepare a cleaning schedule which clearly defines the procedures and roles of all personnel. Implement a regular inspection and auditing system that is documented and insures that action is taken by identified personnel. For example, inspection audits may be done weekly by the supervisor, monthly by the departmental manager, and quarterly by the factory manager.

How to develop a sanitation programme is described in section 4.2.4 with the pheromone monitoring programme driving the cleaning actions.

Plate 4.1 An industrial mobile vacuum cleaner suitable for cleaning in factories.

Plate 4.2 Home-made attachments for vacuum cleaning difficult-to-reach areas. A variety of attachments may be purchased.

4.2.2 STORAGE PRACTICES

Avoid cramped storage. Allow ample space for the product stored with adequate access, light and ventilation. This is normally required by law and fire insurance companies.

Stacks, whether of leaf, rag or finished products must be away from walls or corners, and kept clear of doors, windows and ventilators. Mark out stacking plans with painted lines on the floor leaving at least 60 cm around each tier for cleaning. Leave at least 1 m around tobacco stacks which may require phosphine fumigation to allow placement of the sheet (Chapter 8).

Paint corners and wall–floor junctions white to encourage effective cleaning and facilitate inspections.

Good stock rotation is important; the rule here is **first in–first out (FIFO).**

Store finished products on pallets or shelves. Since pallets themselves are a source of infestation and cross-infestation, inspect and clean them regularly. Adhere to industry standards for pallets.

4.2.3 STRUCTURAL MAINTENANCE

Immediate repair of broken windows, damaged roofs and doors and overall weatherproofing are necessary for good warehousing practice. Doors, windows, fan vents, air bricks and any openings provide pests with easy potential access to buildings. Pipes, drains, cables and ducts passing through walls and floors, must be completely sealed to deny access of all pests.

If a window must be open, an 8-strand/cm screen will prevent insect entry.[62,367] If a door must be open, an air curtain will prevent insect entry. An air curtain is a fan blowing air down the length of the opening, directed outward at a 45° angle, at 40 km/h.

4.2.4 HOW TO DEVELOP A WRITTEN SANITATION PROGRAMME

The development of a sanitation and monitoring programme is shown here, and may be adapted to a stemmery, cigarette manufacturing plant, warehouse complex, by-products processing factory or any factory. Simply follow these procedures and adapt them to your own environment. For example, your environment is a hypothetical cigarette manufacturing factory (Fig. 4.3). Include all areas in your sanitation programme: the grounds, car parks, waste collection/compaction area, and perimeter fenceline. Clearly, areas 4, 8, 9 and 10 (Fig. 4.3) where the tobacco flows are important.

Now let's look at an imagined factory floor(s) (Fig. 4.4). We can see the tobacco flow through receiving, preconditioning, conditioning, pre-blend, casing, drying, flavouring, cutting, blending, cut filler silos, making, and packing areas. To develop our sanitation and monitoring

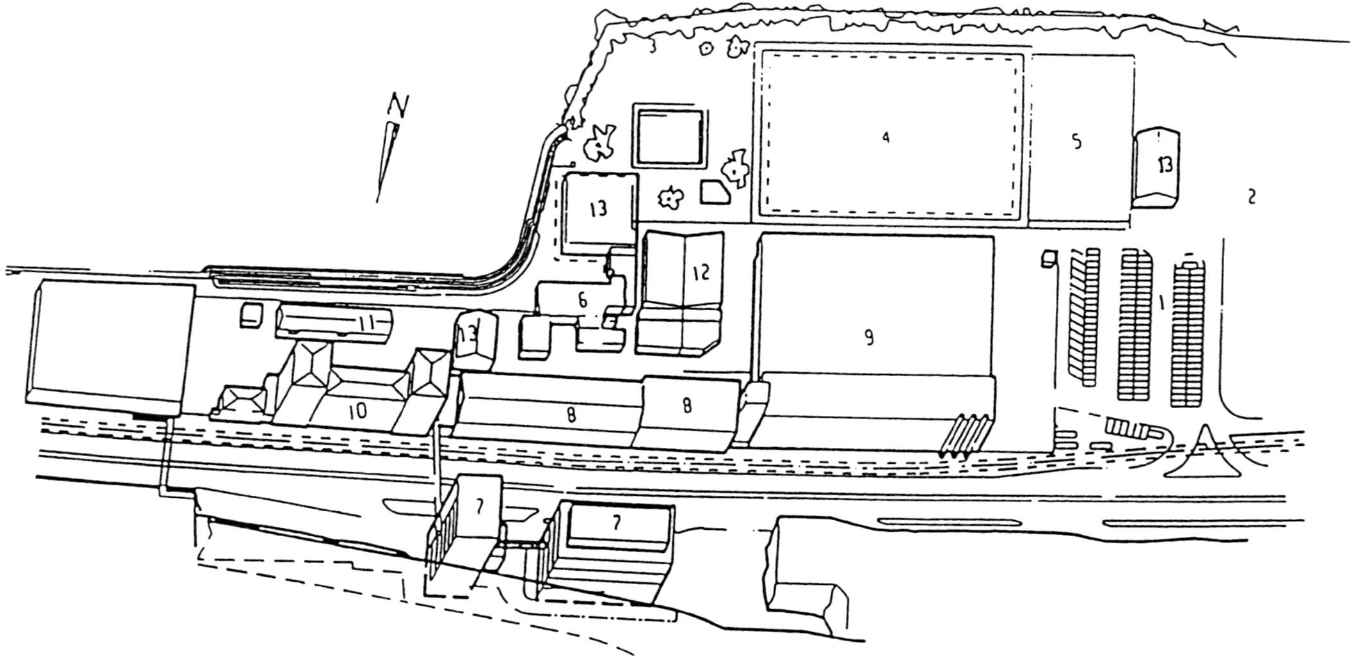

Fig. 4.3 A hypothetical cigarette manufacturing factory. Showing: 1, car parks; 2, lawns; 3, shrubbery; 4, warehouses for leaf and 6, finished goods; 5, boiler plant; 7, administrative offices; 8, 9, 10, the factory proper. The factory is divided into primary, make-pack, and support areas.

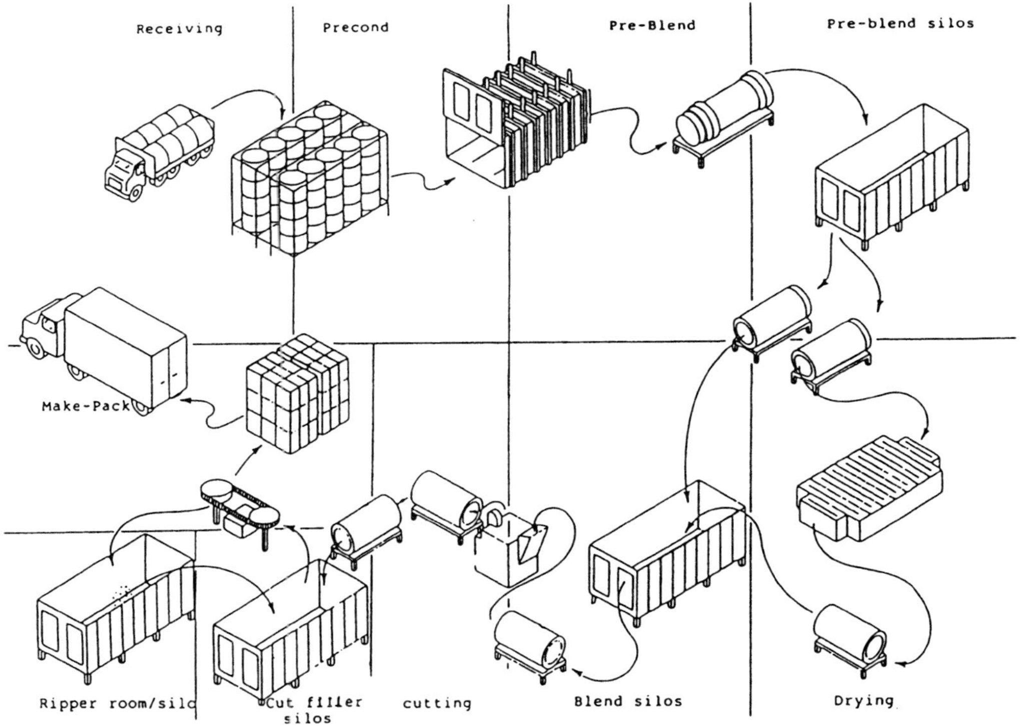

Fig. 4.4 An imaginary factory-floor layout showing the tobacco flow-through: receiving, pre-conditioning, conditioning, pre-blending, casing, drying, flavouring, cutting, blending, cutfiller silos, making, packing, ripping.

programme, break the factory into manageable sections – outside: hard surfaces, lawns, ponds, etc.; inside: lift shafts, underground service ducts; basement storage areas, primary, make-pack, filter making, first floor offices, penthouse service areas, air-conditioning system, dust collector rooms, laboratories. If any of these areas are still too complex, subdivide them – for example, subdivide the primary processing area into conditioning, ordering, blending, cutting/drying, silos, ripper room. This may be arbitrary but usually follows accepted supervisory divisions in your factory. A reasonable example is shown in Fig. 4.5.

Now prepare a schematic diagram for each area or combination of areas. Your engineering department probably already has it. Let us take the ripper room as an example. This is a prime problem area for infestation – and this tobacco is going directly into cigarettes. On your area diagram/plan (Fig. 4.6) mark where your pheromone traps are located (the black • in Fig. 4.6). Do this for every area, where tobacco is present: warehouses, incoming, primary, make-pack, laboratories, finished goods warehouses, penthouses.

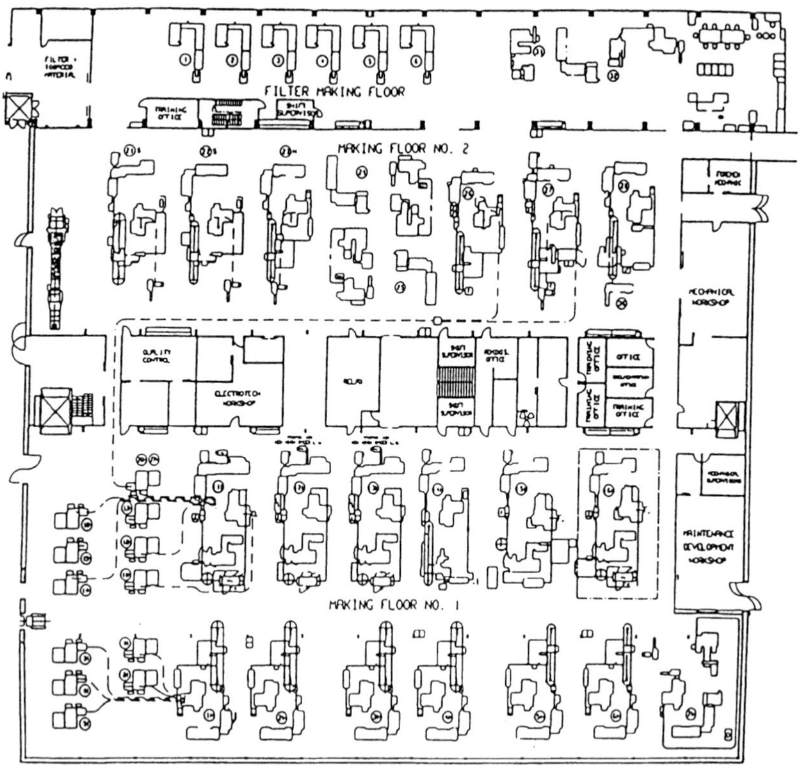

Fig. 4.5 An example of how to subdivide the factory (Fig. 4.4.) into manageable areas for sanitation and monitoring.

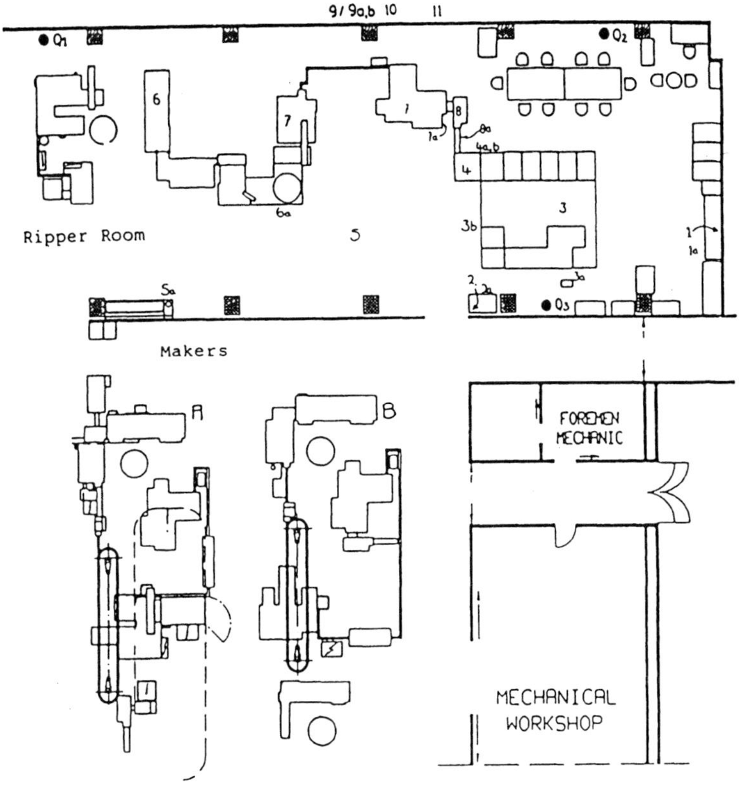

Fig. 4.6 A schematized ripper room showing pheromone trap positions (•). The other numbers correspond to the sanitation report.

Now list each area, and then each trap number and location. You now have a trap report (Fig. 4.7). Simply record the number of beetles caught on each trap each week. This report is crucial – it highlights problem areas for extra cleaning and is used to direct area supervisors' actions (Fig. 4.8). Issue this report quickly to such supervisors. Of course, they should be routinely looking at "their" traps. A monthly summary goes to management. A yearly graphic summary will show problem areas and trends. Include at least 6 months' data on the report, not just a few weeks (only 6 weeks are shown in Fig. 4.7 for simplicity).

Take the same area diagrams (Fig. 4.6) and go with the area supervisors and others such as maintenance, cleaning, engineering personnel and summarize all the places that need cleaning. Yes, of course, that means everything. Then group those places that need daily, weekly, or monthly cleaning. Some may require cleaning less frequently based on experience, such as the roof. Now you probably have written down what you clean

			Number of week /Total beetles captured							
Area	Position	Code	22	23	24	25	26	27	Total	Comments
Total blend										
	Exit Apron 2	H1	0	1	1	0	2	0	4	
	Blend Silo 2	H2	2	0		0	0	0	2	
	Blend Silo 4	H3	0	1	0	1	3	1	6	
	Blend Silo 6	H4	1	2		1	1	2	7	
	Blend Silo 8	H5	0	0		0	0	0	0	
Stem lines										
	Stem Conveyor Y	H6	0	0	0	0	0	0	0	
	Feeder to Belt (01)	H7	1	2	2	2	1	0	8	
	Flattener	H8	1	2	3	2	1	0	9	
	Stem Drier	H9	0	0	4	1	0	0	5	overhead ducts cleaned
	MOBILE TRAPS	MH			8	2	0	0	10	
Cut and dry										
	Cutter # 3	I1	1	1	1	0	0	0	3	
	Cutter # 6	I2	1	2	2	2	0	0	7	
	Conveyor m2	I3	0	0	0	2	2	1	5	
	Drier Exit	I6	0	2	2	2	1	1	8	
	MOBILE TRAPS	MI	–	–	3	3	0	0	6	
Cut silos										
	Silo # A front	J1	0	0	0	0	0	0	0	
	Silo # A rear	J2	0	0	0	1	0	0	1	
	Silo # C front	J3	0	0	0	0	0	0	0	
	Silo # C rear	J4	0	0	0	0	0	0	0	
	Silo # E front	J5	1	0	0	0	0	0	1	
	Silo # E rear	J6	0	0	0	0	0	0	0	
Weekly totals			8	13	16	14	12	3	66	

Fig. 4.7 An example of a pheromone trap report. Areas, or tobacco processing steps, are divided into places and positions, i.e. the exact trap location. Traps may have a code number.

Area	Area code	Persons(s) responsible for area both monitoring & cleaning Shift (A) (B) (C)
LEAF		
Warehouse # 1	A	P. Enderson (A), T. Joke (B)
Warehouse # 2	B	T. Joke (B)
Sample Room	C	F. Lake (B)
Laydown	D	etc.
PRIMARY		
Conditioning	E	
Ordering	F	
Pre-blend	G	
Total Blend	H	
Cutting/Dry	I	
Cut Filler Silos 1-6	J	
Cut Filler Silos 6-10	K	
Special Brand Line	L	
Dust Collectors	T	
Makers B	U	
Filter Making	V	
Dust Collection	W	
QA Labs	X	
FINISHED GOODS		
Warehouse	Y	

Fig. 4.8 A check-list by area of supervisors responsible for monitoring and sanitation.

already – what we may refer to as *general* cleaning (Fig. 4.9). Annotating the area plan(s) is useful to visualize everything, avoid missing places, and cut down on paperwork. If these diagrams may be computerised, even better, especially when modifications are made.

Now, detail how each place is to be cleaned. This written programme is also useful for training new employees in cleaning and safety procedures such as locking equipment prior to cleaning or isolating electrical devices. Now you have the basis of a written cleaning programme. We want to add the thorough cleaning that *prevents* tobacco accumulation and thus infestation. We call this, thorough, or in depth cleaning, or "smart" cleaning, because it outsmarts the beetles!

Area	(Code)	Place	Number	Frequency (D) daily, (W) weekly (M) monthly - other	Method used
Ripper Room	(Q)	Behind electrical panel	1	M	Vacuum + attachment 8cm
		Behind electrical panel	2	M	Vacuum + attachment 2cm
		Ripper Silo Inside	3	W	Vacuum – see meth. #40 lockout
		Conveyor to silo	4	M	Vacuum + attachment 10cm
		Floors	5	D	Vac/Broom
		Ripper feed table	6	D	Wipe cloth
		Ripper body	7	M	Vacuum-see meth. #45 lockout
		Ripper conveyor outlet	8	D	Vacuum + attachment
		Ceiling	9	3 mos	Forklift + Vacuum
		Overhead conduit	10	M	Vacuum + attachment (ʃ) 10cm + brush
		Above false ceiling	11	Yrly	Vacuum + sweep attachment

Fig. 4.9 A general cleaning programme for a ripper room (Fig. 4.6) as an example. Numbers refer to locations in Fig. 4.6.

Area	(Code)	Place	Number	Frequency (D) daily, (W) weekly (M) monthly	Method used
Ripper Room	(Q)	Behind electrical panel	1	M	Vacuum + attachment ╗ 8cm
		False base of panel	1a	M	Vacuum
		Behind electrical panel	2	M	Vacuum + attachment ╗ 2cm
		False base of panel	2a	M	Vacuum
		Ripper silo inside	3	W	Vacuum - see meth. #40 lockout
		Under plate of ground duct	3a	M	Vacuum + attachment ╡| 10cm
		Angle (∪) underside of silo	3b	M	Vacuum + attachment || 4cm
		Conveyor to silo	4	M	Vacuum + attachment ╡| 10cm
		Under selvage	4a	W	Vacuum + hook (J) attachment
		False sides panels	4b	M	Vacuum + flexible hose
		Floors	5	D	Vac/Broom
		Drains	5a	M	Steam hose and wire brush
		Ripper feed table	6	D	Wipe cloth
		Ledge underneath roller	6a	M	Vacuum + inverted (J) attachment
		Ripper body	7	M	Vacuum-see meth. #45 lockout
		Space below machine	7a	M	Vacuum + knife attachment
		Ripper conveyor outlet	8	D	Vacuum + attachment ╥ 10cm
		Closed duct above support	8a	M	Vacuum + attachment ═ 12cm
		Ceiling	9	3 mos	Forklift + Vacuum
		Inside light fixtures	9a	3 mos	Cloth
		Inside roller door fixture	9b	3 mos	Vacuum
		Overhead conduit	10	M	Vacuum + attachment (ſ) 10cm + brush
		Above false ceiling	11	Yrly	Vacuum + sweep attachment

Fig. 4.10 A written smart cleaning programme based on Fig. 4.6, a ripper room. This upgrades Fig. 4.9's general programme to include additional insect harbourages.

Area	Code	Place#	Frequency	Date cleaned (wk)	Maintenance modifications	Comments + inspector's initials
Ripper Room	Q	1	M	3/9 4/7 5/6 6/8 7/3	Fast release on base plate	Beetles found 5/6
		1a	M			high trap count
		2	M			
		2a	M			
		3	W			
		3a	M			
		3b	M			
		4	M			
		4a	W			
		4b	M			
		5	D			
		5a	M			
		6	D			
		6a	M			
		7	M			
		7a	M			
		8	D			
		8a	M			

Fig. 4.11 An example of the smart cleaning programme used to verify and inspect, as well as direct, maintenance work.

Go back to the same area (Fig. 4.6) and take another look, lie on the floor, climb up high, and get under and inside equipment (safety first). As you do this use a **torch** (flashlight) with a powerful beam to look for all the places tobacco has accumulated. Do this after the general cleaning. When you see all the tobacco that remains, you will know what smart cleaning involves. Now list all those places and include them on your written programme, with instructions on how to clean (Fig. 4.10). Many of these places will require adaptors for the vacuum cleaners to get into awkward places. Buy or have the adaptors made: they are a good investment (Plate 4.2). Also note all the places that were hard to find because you had to unscrew/unbolt plates, etc. These places are listed so maintenance can alter them to allow rapid access and/or prevent their accumulating dust (Fig. 4.11).

With this done, you have a written sanitation/monitoring programme. Now you *only* have to implement it. Start with meetings with personnel by area and/or shift. Explain the need for this and how it is going to work and their role. Get their acceptance for smart cleaning and **vacuum** cleaning. Train a good cleaning crew and then use them to demonstrate techniques.

Your written programme can easily be used to verify that places have been cleaned. A copy may be left in each area and personnel are to sign when places have been cleaned with the date (Fig. 4.11). Your role and that of other sanitarians or consultants will be to check (at random) that places said to be clean really are. A good example is to inspect a cigarette maker after its monthly smart clean. It should be spotless. Since you know how effective the torch is, provide plenty for each area so that places get cleaned properly.

With all this done, and yes it will take time, you are ready to fine-tune or focus. Use pheromone trap counts to direct attention to problem places. Then use mobile traps and your eyes (with the torch!) to locate these "new" harbourages. Add them to your written programme. As the programme evolves, so your sanitation standards improve. Whenever you can, get "a new pair of eyes" to audit your programme and make suggestions for improvements.

Your goals are primarily no *L. serricorne* detected in your monitoring system, and, secondarily a fast, responsive sanitation programme to remedy any insect problems found. To develop a programme like this and to ensure it improves requires someone to be responsible. This person must be able to coordinate the efforts of cleaning, production, maintenance, grounds, warehouse, safety, quality control and management personnel. Before you start on this project, make sure you have read this section and understand it. Now call a meeting and persuade the management team, including the plant manager (or equivalent). They must be *convinced* and *committed* to helping you succeed in developing an excellent sanitation programme.

Experience shows the initial costs and effort are amply repaid by reduced, efficient cleaning efforts (and costs) later. An example would be: a factory producing 40 billion units each year would spend at least $2 million a year on sanitation. An investment of 10% or $200,000 over 2 years in equipment, and development of a **Factory Sanitation Programme** will result in reduced costs of at least 10% or $200,000 every year thereafter and vastly improved sanitation.

4.3 SPECIFICS DURING THE TOBACCO PROCESS (FIG. 1.1)

4.3.1 ON-FARM

Infestation may begin post-curing. Carry-over or tobacco stored on-farm poses a great risk of infestation.[205] Rapid transfer from the farm to processing is preferred. If the farmer must store tobacco, it must be in clean storage barns with screened windows and doors (Plate 4.3). A pheromone trap should monitor for any potential infestation. Industry groups and extension agents must stress the need for sanitation on farms. On-farm storage, before marketing and during one or more years' "carry over", in infested barns results in massive infestations entering the leaf supply channels. **Improved sanitation during on-farm storage and pre-market storage is a long overdue requirement and cannot be over-emphasised.**

Plate 4.3 On-farm storage of tobacco with no preventive measures leaves tobacco vulnerable to infestation.

4.3.2 PROCESSORS

Sanitation, as a preventive or control method, is seldom practised vigorously in tobacco processing plants. During the season when the processing plant is in full operation, little is done to remove tobacco residues other than to blow or sweep visible surfaces, floors, and machinery. No in-depth cleaning is done until the final tobacco is packed and has left the plant. Indigenous insect pests that come in on the crop (see above) are likely to establish infestations throughout the processing plant, thereby creating reservoirs of infestation for tobacco being prepared for shipment. Good sanitation during the processing season is essential to prevent insect problems being passed on to leaf in storage. The Industry should influence packing stations to improve sanitation, and develop sanitation programmes as outlined above (section 4.2.4) for manufacturing factories.

4.3.3 SHIPMENTS

The buyer or importing country often request fumigation of a tobacco purchase, prior to shipment, as a necessary precaution against the ubiquitous tobacco pests. Sound containers protect tobacco [and may serve as ready-made fumigation chambers (Chapter 8)]. Nevertheless, there should be no evidence of the previous cargo, with containers free of taint or odour. Clean containers thoroughly before use with a brush and vacuum cleaner. An additional monitoring precaution would be to tape a pheromone trap on top of a case at the back of the container for examination at the destination warehouse. Re-infestation in-transit is a distinct possibility given the presence of *E. elutella* in many towns[367] and *L. serricorne* in dockside warehouses.[264] Therefore, the risk of beetle entry via container vents or during customs inspections should be considered and as a precaution, the vents screened.

4.3.4 WAREHOUSING

Segregate incoming tobacco shipments until inspected and fumigated if necessary, and then move it into the main warehouse. Inspection regimes vary widely, but there is a recommended minimum of cases to be visually inspected (section 3.3). This minimum depends on the origin of the tobacco and the possibility of infestation. For example: "*Non-suspect*", (3%); "*Suspect*", 10%; "*Special Purchases*", fumigate on receipt and then inspect. This arbitrary system depends on the infestation history of a given supplier. Communication of information between the Industry's buyers, shippers and warehousemen ensures vigilance regarding "suspect" tobacco.

4.3.5 FACTORIES

The diversity of activity here means that every location and process will be inherently different. The common theme in manufacture is the establishment of a sanitation standard with documented procedures for both sanitation and inspection programmes which incorporate a feedback or corrective loop system (Fig. 4.2).

The treatment and handling of trade returns usually takes place at the original point of manufacture. As these may be a source of re-introduction of infestation to the factory, adopt a system of segregation and disinfestation (or destruction). Various physical and chemical methods of disinfestation are available (Chapters 5 and 8), including cold treatment deep freezing which is used by a number of companies.

A list of suggestions for preventive sanitation is given below; these concern plant layout, equipment and design. Adherence to these points will reduce the amount of cleaning needed and, therefore, costs.

Landscaping and grounds

1. Gravel or concrete perimeter around the factory > 1 m.
2. Shrubs and trees >5 m from the factory, preferably only lawn and flower beds.
3. Trash collection area: >10 m from factory, walled area, concrete base, drained .
4. Exterior lights away from, but aimed at, work areas: sodium vapour type.
5. Perimeter lights >20 m from factory: mercury vapour type or sodium.
6. Other outside lights (e.g. doorways): low-watt yellow.
7. Good drainage to avoid any standing water and accessible for cleaning.
8. A >2 m clear border outside the perimeter fence.
9. Factory air intakes with exterior, 3 m diameter, hemispherical cordon.

Factory construction

1. Access and space is available for cleaning activities.
2. Wall/floor junctions are curved ╰ and not └-shaped: no skirting boards.
3. All areas are accessible by vacuum cleaner.
4. Walls, ceilings, floors, and equipment are light-coloured: cream or grey.
5. Maintenance, repair, and installation work is finished: no unfinished holes.
6. Adequate lighting, use secondary standard in primary and warehouses.
7. No false ceilings or false walls (wall voids).

8. No underground ducting, drains accessible, flushable, and rodable.
9. Separation between pre-direct conditioning cylinder (DCC) | primary | cut filler storage | secondary.
10. Dust collection area is separated from the factory.
11. Ripper room is segregated, in a separate room or a screened (8/cm) area.
12. All openings are screened, e.g. windows 8/cm screen; doors + air curtains.
13. Construction is simple and allows easy access.
14. Removeable mesh traps are present on all drains, inside and outside.
15. Walkway access to all overhead equipment, conveyors and pipework.
16. No **storage** of anything in production areas.
17. Exit doors of good fit, self-closing, with sensor to detect if propped open.
18. Roll doors with flexible bottom "seal" and T extensions to fit rail tracks.
19. Design out ledges of any kind, this | not, ˥, ⊣, or ⊥, inside or outside the factory.
20. Structural metal work is not enclosed, **H** beams with sloped ∠ inner base.
21. No sunken areas, except wet cleaning catchment areas.
22. Crawl spaces with adequate walkways and space.
23. All walls are smooth inside and outside, no rough brickwork, etc.
24. Expansion joints sealed flush with the floor.
25. Gas-tight construction for fumigation rooms, attention to wall/floor/roof junction.
26. No food allowed on-site during construction: crew canteen at perimeter.
27. Kitchen and restrooms separate from production area by two sets of doors.
28. Restrooms and food preparation areas are completely tiled.
29. Restrooms with showers and foot-operated water and soap and towel dispenser.
30. Lift shafts drained.

Equipment design and installation

1. Suitable vacuum attachments are available in adequate numbers.
2. All installations are: >30 cm from the wall, such as electrical cabinets.
3. >30 cm from the ceiling, such as silo shuttles.
4. >30 cm off the floor, not <10 cm, this includes shelves.
5. No cupboards or cabinets in the factory. Use peg-board if essential.
6. No dead-end ducting or pipework.

7. All metalwork must be ∩, O, Λ, or | - shaped, not ∪, V, ⊥, or J - shaped.
8. Cabletrays have open bases, with 1 cm between cables.
9. No "hermetic" sealing, there must be access, hermetic does not last long.
10. Equipment and installations have no "voids".
11. Access to "self-cleaning" dust extractors, conveyor catch pans, etc.
12. Weld closed hollow tubing, or removable plugs installed.
13. Covers may be removed from covered cable trays.
14. Design for maximum "drop through" of tobacco to the floor (e.g. see 15 and 16).
15. Chain guards and motor housings with meshed base.
16. Conveyor catch pans avoided or hinged and 20 cm below the conveyor.
17. Access to, and able to see, inside of all equipment.
18. No double-walled equipment.
19. Equipment, especially makers, have no hidden, inaccessible voids.
20. Require equipment suppliers to provide **detailed** machine cleaning specs.
21. No skirts to hide the gap between the equipment base and the floor.
22. Overhead mirrors or light fixtures with ∩ shape, not ∪.

4.3.6 MACHINERY

Whenever machinery is moved the machine must be stripped, thoroughly cleaned, and infestation free. This may be checked visually and with pheromone traps. With all transfers and purchases, the despatching location guarantees this cleaning.

4.3.7 DISTRIBUTION TO CONSUMERS

A temperate climate, several layers of packaging, and a relatively short shelf-life usually protect the finished product throughout the distribution and retail chain. The likelihood of infestation is slight, but sanitation, inspections and good storage practices are necessary. Vending machines are often exposed to feral beetles and therefore, must have monthly stock rotation.

Products going to tropical markets have long journey times from the place of manufacture. Layers of packaging may no longer be a protective barrier in countries where the cigarette beetle is indigenous. Inspection, sanitation and storage practice have greater importance and are supported by other preventive measures, such as chilled storage.

Issue guidance to all distribution and retail personnel outlining the problems, the tobacco pests and the importance of preventive measures, e.g. sanitation and inspection.

4.4 NON-TOBACCO EATING PESTS

4.4.1 INSECTS

A good sanitation programme will also prevent non-tobacco pest infestation such as outlined below:

Wood-boring insects may reduce wood and pallets to dust causing tiers to collapse. Adult insects when emerging from wooden pallets will continue to bore through packaging and into the tobacco if their way is blocked. Look for holes in the wood or piles of powdery wood dust underneath pallets. Common findings are: tropical bamboo or other borers (Bostrichidae), the powder post beetle (*Lyctus bruneus*), and horntails (*Urocercus gigas*).

The mould-eating hairy cellar beetle, *Mycetaea subterranea;* fungus-eating hairy fungus beetle, *Typhaea stercorea;* and dried foodstuff-eating drugstore beetle, *Stegobium paneceum,* may be found in warehouses, and perhaps confused with *L. serricorne* (section 2.3). Their presence indicates poor storage conditions. Remedial action requires improved housekeeping and/or improved ventilation/heating.

Termites, whether subterranean or drywood, cause considerable damage to buildings. Their presence in a building requires specific treatment of the building or replacement by pre-treated materials and foundations treatments to provide an insecticidal barrier between soil and the building. Consult a specialist and experienced contractor.

The mite species *Tophagus* is rare and associated with mould outbreaks on which the mites feed. The higher moisture levels of pipe tobaccos, poor sanitation and suitable atmospheric conditions lead to fungal blooms and consequent population explosions of mites. Mould spoilage of leaf tobacco due to water damage in transit or storage can similarly lead to outbreaks of mites if the damage is not detected early enough. These mites may cause skin reactions.

Other insects may cause nuisance to employees or may cause marking or spoilage of packaging, although few will enter the product. Cockroaches, for example, will leave faecal smears on packaging. Good sanitation will deter such pests. Check out any suspicious insect findings. Insects imported with the leaf are generally due to cross-infestation from other cargos. These are relatively rare now due to containerisation, but infestation of *Tribolium* species may occur. Other insects may be found in containers as a result of infestation of a previous cargo and poor sanitation. If these insects are still alive then the efficiency of the tobacco leaf fumigation should be questioned. *Ephestia kuehniella* and *Plodia interpunctella* are common finds in poorly cleaned containers.

Finally, cockroaches may enter opened cigarette packs in consumers' homes. Returned cigarettes, usually from warm climate areas, will show faecal and oothecal smears, bite marks on the paper and, occasionally, the presence of egg sacs (ootheca) (see section 9.7).

4.4.2 VERTEBRATE PESTS

Rodents pose an obvious adverse health risk and nuisance value to employees. Both rats and mice may cause considerable damage by gnawing. Most large premises have some form of regular rodent monitoring. Look for signs of gnawing and excreta. Consider the possibility of contamination from toxic rodent baits. Should such bait-stations be used, they must be tamperproof and preferably mechanical. Do not allow cats to roam through warehouses. Their excreta and urine can cause damage and taint problems. Cats are not an acceptable method of rodent control.

Nesting or roosting birds may cause a range of problems. Droppings are highly acidic and may damage all surfaces. Dry, dusty droppings may cause chest complaints and breathing difficulties. The finished product is at most risk from taint due to bird droppings. Many insect pests are associated with birds, parasitizing the living birds, colonising the nests, living off dead birds, feathers and even the droppings. With a correct sanitation standard, vertebrate pests will cease to be a problem. Of course, screening is an important means to prevent access by birds.

Physical control methods 5

Infested tobacco may be treated effectively by fumigation (Chapter 8); however, there are occasions when this is not possible due to local regulations, logistics, or the ambient temperature. Some alternative methods are discussed below, along with their limitations.

5.1 TEMPERATURE

One must stress that the insects are usually insulated against temperature changes by the mass of tobacco.[300] To guarantee complete disinfestation, the time required to heat up or cool down the tobacco has to be added to the time necessary to kill the insects. This time is often far in excess of the actual insect killing time. The use of temperature-monitoring devices is necessary to ensure the centre of the treated tobacco reaches the killing temperature for the required duration.

5.1.1 COLD

In the reviewed literature dealing with physical means to control infestation, low temperature, or deep freezing, seems to be the most promising method. However, it has to be taken into consideration that the *E. elutella* is reported to be little affected by low temperature.

Diapausing *E. elutella* larvae may survive exposure to –18°C for several days. They may also survive in unheated buildings, where the winter season ambient temperatures are as low as –34°C.[342]

Nevertheless, if attention is paid to the type and volume of the insect surroundings (e.g. bales, hogsheads, finished products in cases on pallets), satisfactory results in killing *L. serricorne* and *E. elutella*[42] may be obtained.[173] Indeed, the published requirements of the temperature and duration required to achieve 100% insect kill are summarized in Table 5.1.

The results (Table 5.1) have led to two cold treatments to kill 100% of *L. serricorne*: **Freezing**, where tobacco is exposed to less than –20°C for greater than 5 days.[351] This may be accomplished in commercial, bulk, draft-assisted freezers or using liquid air.[336]

Cold storage of tobacco at less than 4°C for greater than 90 days. This is commonly used for cigarette buffer stocks and in limited instances for green

Table 5.1. The duration of cold temperature required to kill 100% of *L. serricorne* or *E. elutella*

Insect*	Surrounding temperature (°C)	Exposure time
Exposed insects: treatment resulting in 100% mortality		
L. serricorne		
E, L, P, A	–9	3 days[333]
	–1	14 days[333]
E	–20	2 hours[254]
L	–12	11 hours[283]
E. elutella		
E	7.5	21 days[31]
E. elutella		
E, L	6	>35 days[328]
Insects in infested tobacco: treatment resulting in 100% mortality		
L. serricorne, E, L, P, A		
B‡	–11	8.5 days[333]
B	–23	5 days[342]
B	–10	28 days[297]
C	–23	4 days[341]
H	4.5	12 weeks[92]
H	9	32 weeks[92]
H	4.5	91 days[143]

*Insect treated included: eggs (E); larvae (L); pupae (P); adults (A).
‡Tobacco packed in: bales (B); cases (C); hogsheads (H).

leaf stock. Possibilities exist for designing warehouses to take advantage of cooling to inhibit or control infestations[27,28,86] as in the grain industry.[61]

A survey of various methods of infestation control of tobacco used in the tobacco industry was conducted in 1975[107] and included cold treatment. Below are examples of cold treatments of various types of packed tobacco or finished products, that will result in 100% *L. serricorne*/*E. elutella* mortality, commonly used in our industry.

Samples

Before examination of incoming leaf samples, they should be placed in deep freezers (below –20°C) for 5 days. Samples should be 10 cm apart to allow air circulation. Keep samples in sealed plastic wrappers. This technique may be used for other small samples of processed/manufactured tobacco. It has been shown that 24 hours at –20°C is sufficient.[193] The additional 4 days are suggested to err on caution and ensure kill of any diapausing (overwintering) immature stages of *L. serricorne* or *E. elutella.*

Manufactured tobacco

Cut tobacco in cases (1 m^3, 80 kg) or cigarettes (200/carton; 50 cartons/case; 24 cases/pallet) may be placed in a commercial draught-assisted freezer (–30°C) for 5 days. Tobacco or cigarettes should be reconditioned or tempered for 4 days. Such treatments show no adverse quality effects.

5.1.2 HEAT

Published reports on the effect of heat on *L. serricorne* are summarized in Table 5.2.

Clearly, tobacco containing *L. serricorne* has to be treated longer at comparable temperatures than the isolated insects due to tobacco acting as an insulator.[300] Eggs are reported to be most heat-tolerant.[275] *E. elutella* is little studied: one report indicates 16 hours at 45°C is necessary to kill diapausing larvae.[38] Heat is used to condition the leaf prior to processing for pliability and machinability, and again during post-processing drying to reach the correct moisture for manufacture or storage. The difference in duration/temperature between drying and conditioning may be attributed to the steam or wet heat used in conditioning resulting in faster kill. Heat is used as a safeguard since processing variability often results in failure to meet, exactly, the necessary killing temperatures. However, these two processing steps may be regulated and monitored to ensure insect kill.[262] Indeed, conformance to an insect killing standard is recommended: it is a free "side effect" from normal tobacco processing.

Table 5.2 The duration of high temperature required to kill 100% of *L. serricorne*[173]

Temperature (°C)	Duration (minutes)	Remarks and life stages tested
No, or minimal insulation from tobacco		
83	15	Eggs on leaf
77	6	All stages, 100% RH
67	45	All stages on leaf
65	5	All stages
60	5	Eggs, 75–80% RH
60	15	Adults, 75–80% RH
Insulated within tobacco mass		
60	105	Within bundles leaf
60	30	Eggs and larvae in cigars
60	60	Pupae in cigars
59	120	All stages in boxed leaf
57	300	All stages in bales
50	300	Bales in air at 60°C, 65% RH
44	600	All stages in bales[275]

Vacuum-steam conditioning parameters commonly used in the industry are based on work with exposed insects, i.e. no tobacco present[26,76,77] and industry trials with infested tobacco. An example of such parameters is: flowing steam at 76°C for 15 minutes under 584 kPa vacuum.[352] The time taken to evacuate the conditioning chamber and additional reconditioning cycles will extend the process to 40 minutes or more. The use of steel probes inserted into the tobacco mass to pull the vacuum from within will shorten the necessary cycle. Given the variable chamber design and vacuum pump efficiency, each system should be tested. This may be done simply by monitoring exit tobacco temperature or via remote temperature recorders housed within the chamber. Clearly, direct cylinder conditioning, without vacuum, will be different. However, the absence of vacuum is offset by the open aspect of the tobacco. Exit tobacco temperature and residence time in the cylinder will indicate if insect-kill is likely. Numerous, unpublished, industry trials have shown that a "steamed" tobacco temperature of at least 60°C for 3 minutes (or 70°C for

2 minutes or 55°C for 4 minutes) will kill all life stages of both *L. serricorne* and *E. elutella*.

Redrying is the process by which the moisture content of tobacco is regulated as preparation for storage. In redrying the tobacco is passed through a long chamber and heated to drive off moisture. Then, steam is introduced to supply the amount of moisture required. Redrying of leaf tobacco with a redryer having a 43-minute cycle causes 100% mortality of all stages of *E. elutella* and *L. serricorne*. Most of this mortality occurs in the drying chamber. Redried tobacco strips, therefore, will *probably* have no insect infestation, but it is possible that a few insects may survive the process.[340] This was confirmed during tests of three redrying plants for strips.[144] Unfortunately, a modern trend is toward short operating cycles (5 to 6.5 minutes) which exposes insects to temperatures higher than 60°C for only 2 minutes. Under such redrying conditions, mortality of all the stages of *L. serricorne* ranged between 75% to 99%.

Most insects that survive primary processing are likely to be killed in the **dryers** prior to cut rag storage. Dryer temperature and duration are specific to companies and blends; thus consult the dryer specification and dryer temperature/duration profile. Insects at this stage may be considered exposed therefore to a tobacco temperature greater than 60°C for greater than 3 minutes, which may be expected to kill all insects.

Microwaves have been suggested as an alternative to conventional heating procedures, due to the faster transport of energy and therefore, shorter duration required to deliver a given dose.

Microwave heating at 55±2°C of cut tobacco, leaving a rotary dryer resulted in 100% mortality of all stages of the *L. serricorne* and the *E. elutella*.[171] However, this method is not feasible, because the initial tobacco temperature must be high, otherwise the energy demand to obtain killing temperatures in a short duration is costly. The lethal effect of microwaves is the effect of heating, which may be efficiently achieved conventionally during the manufacturing process.

5.2 RADIATION

Disinfestation and decontamination of some foods, spices, etc. by gamma-irradiation is widely used in many countries. However, there are severe restrictions in other countries (e.g. Germany). The effect of radiation is based on photochemical reactions and on the generation of radicals, ions and their reaction products. These interfere with the metabolism of living cells, thus leading to cell death. Radiation procedures are expensive and generally require stationary equipment and extensive safety precautions. It is likely that where irradiation has been used, packaging must be so labelled. The effect of such irradiation on tobacco quality characteristics is unstudied. Indeed, radiation, although researched, is not presently used by the tobacco industry.

Gamma-rays are electromagnetic rays of high energy, wavelength 10^{-12} to 10^{-14} m, emitted by radioactive elements, such as cobalt-60. Gamma-irradiation is effective in killing insects (disinfestation) or microorganisms (decontamination). Doses required for decontamination usually are higher than those required for disinfestation. Adult insects are the most tolerant, followed by pupae, then larvae and eggs.[345]

The published results of some of gamma-irradiation tests are summarized in Table 5.3.The high radiation doses necessary to kill adult *L. serricorne* are similar to those used to decontaminate spices (7 kGy).

X-rays are similar to gamma-rays, but have less energy and wavelengths between 10^{-10} and 10^{-12} m. As early as 1916, under laboratory conditions, exposure of tobacco containing all stages of *L. serricorne* to high-energy X-rays stopped the activity and development of larvae, while adults survived as long as the controls. Deposited eggs of the irradiated beetles were infertile, as were the adults from transformed irradiated pupae.[298]

It seems that X-rays may not be appropriate for a general practice of disinfestation of tobacco. However, with similar doses, X-rays are more effective than gamma-rays from cobalt-60 radiation, due to higher "relative biological effectiveness".[365]

Table 5.3 Effects of gamma-irradiation on *L. serricorne* and *E. elutella*.

Insect*	Dose (Gy)	Result
L. serricorne		
P	100–360	Sterility only[270]
P	60–80	Sterility[71]
P	500	100% mortality[71]
A	130	Sterility[71]
A	1500	99% mortality[71]
E, L, P, A	250	Sterility[345]
E, L, P, A	1000	<100%[345]
E, L, P	750	100% mortality[128]
A	2500	100% mortality[128]
L, P, A in bales	600	13% survival[323]
E. elutella		
A (1 day old)	550	99% sterility[59]

*Insect treated included: eggs (E); larva (L); pupa (P); adult (A).

5.3 PRESSURE

High pressure carbon dioxide at 4,000 kPa for 30 minutes[157,158] or 3,000 kPa for 50 minutes[158] kills *L. serricorne*[324] and is accepted as a tobacco treatment by the German authorities, although not yet used in the industry. This "Carvex" process[157] may be used on all tobacco – with caution – as it may also affect cigarette pack materials. The effect on tobacco quality characteristics is not known (see also fumigation in section 8.5).

Low pressure requires long duration for adequate mortality. Suitable treatments would be 100 kPa for 10 days to kill all *L. serricorne* or 3 days for *E. elutella*.[24] Lesser regimes are known to have little effect.[258] Studies with vacuum packaging showed *L. serricorne* to be low-pressure tolerant for several weeks.[168]

5.4 PHYSICAL BARRIERS

Physical barriers as packaging may help prevent re-infestation of stored tobacco or tobacco products. *L. serricorne* have powerful mandibles (Plate 5.1), which enable the insects to attack and penetrate packaging

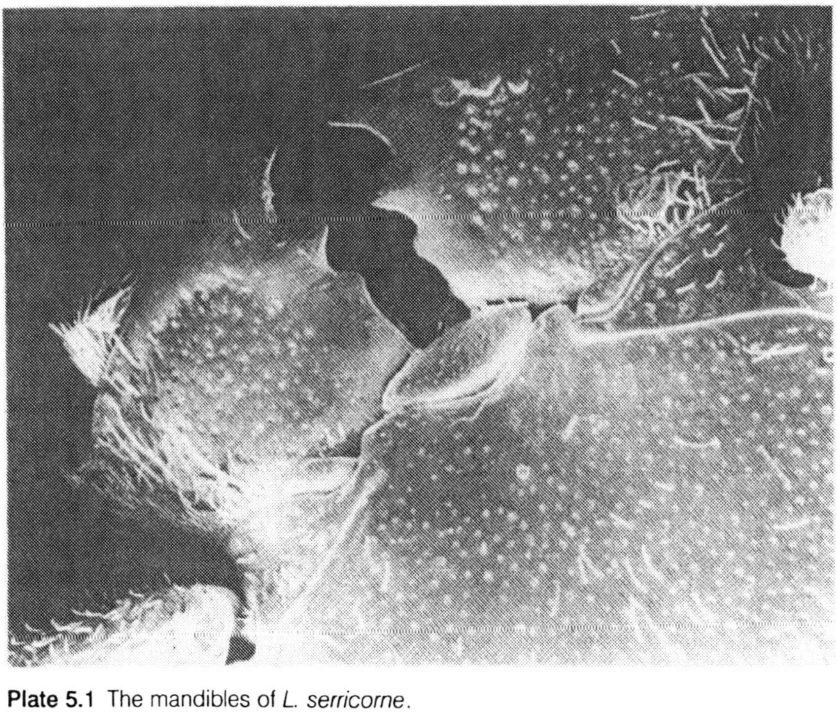

Plate 5.1 The mandibles of *L. serricorne*.

materials. Their ability to penetrate various packaging materials has been published.[197,198] Some of the results of papers studying the effectiveness of physical barriers are summarized in Table 5.4.

Of course, screens are useful physical barriers against insect, rodent and bird entry to buildings.

Good physical barriers include: 25-μm high-density polycarbonate; 38-μm high-density polypropylene; 100-μm low-density polypropylene; 200-μm spunbonded polyethylene.[137] Polypropylene is more suitable than cellophane.[169]

Table 5.4 Resistance of packaging materials to *L. serricorne*

Experiment:	Laboratory screening[137]
Material tested:	High-density polycarbonate 0.0254 mm
	High-density polypropylene 0.0381 mm
	Low-density polypropylene 0.1016 mm
	Spunbonded polyethylene 0.2032 mm
Result:	All good physical barriers
Experiment:	Cases with liners stored for 1 to 2 years in an infested warehouse[137]
Material tested:	See above
	+ spunbonded polyethylene 0.2540 mm
Result:	No adequate sealing, all cases infested, best result with spun-bonded polyethylene
Experiment:	Bales with overwrap stored for 1 year in an infested room[137]
Material tested:	Spunbonded polyethylene
Result:	0.2032 mm most promising if sealed adequately
Experiment:	Field trial; cardbord cases with liners stored 495 days[362]
Material tested:	Spunbonded polyethylene 0.1270 mm
Result:	5% of the cases penetrated versus 52% controls
Experiment:	Packages of cigarettes with overwrap placed in infested chambers for 1, 2 and 6 months[169]
Material tested:	Polypropylene 0.0191 mm versus cellophane
Result:	Polypropylene far more resistant to insect penetration than cellophane

The barriers are only as good as their joint seals; *L. serricorne* will enter via poor seals.[137,362] The small first instar larvae are able to penetrate cigarette pack outerwraps when poorly sealed. Often *L. serricorne* is thought to have eaten its way into tobacco packaging, whereas, really the first instar larva crawled inside and the emerging adult or larva created a hole in escaping (section 2.1.1 and 9.7)!

5.5 LOGISTICS

Segregation of tobacco on the basis of type or age contributes to controlling the spread of *L. serricorne* or *E. elutella* infestation within a warehouse complex. Segregate newly purchased tobacco from "clean" tobacco if the infestation status of the former is not known or is suspect.[233] Because tobacco is often held for hours or even days before it is processed, do not bring infested tobacco into the manufacturing plant.[350]

In the factory, overstocked tobacco should be avoided because it is a potential source of infestation. Tobacco stocks should be processed following the first in–first out (FIFO) principle. Storage time in the factory should not exceed 1 week and cases with scrap tobacco should be removed. These considerations form part of the sanitation standard (Chapter 4).

6 Insecticides

Where a comprehensive sanitation programme is not in place (Chapter 4), or where it is not rigorously enforced, tobacco and the factory may become infested. In such cases insecticides may be necessary to protect or disinfest the tobacco during periods of long-term storage at the country of origin, in transit, at the tobacco manufacturers' warehouses or production facilities. Infestation risk to tobacco occurs in tropical and temperate areas where facilities, especially those for production, are heated. Feral reservoirs of tobacco insects are fewer in temperate areas.

Insecticidal treatments for tobacco protection include insect growth regulator application (Chapter 7) and fumigation (Chapter 8). Space spraying [244] and residual treatments of building surfaces to offer effective knock-down and kill of adult insects is also used.[6,256] Fumigation rapidly controls existing infestations; however, it provides no residual insecticidal activity. Alternatively, chemicals like methoprene applied directly to tobacco provide long-term insecticidal protection, but not rapid control.

We discuss, below, the use of conventional insecticides as space and surface sprays. In most countries such insecticides are controlled substances, as indeed are fumigants and insect growth regulators (IGRs). Persons applying them must be trained in their use and, often, licensed by appropriate government agencies. **The insecticide label and instruction manual must be followed**, along with other published guidelines, to ensure responsible use.[109]

6.1 MODE OF ACTION

Space and surface sprays are contact poisons with the exception of the IGRs (Chapter 7). They are particulate and their activity depends on direct contact with and penetration of the insect cuticle. These conventional insecticides, such as natural pyrethrins or synthetic pyrethroids, are insect nerve poisons that target, and quickly kill the adult insect. IGRs, such as methoprene, disrupt the insects' life cycle. Methoprene is eaten and prevents maturation of the grown larva to a viable adult (Chapter 7). Insecticidal fumigants are gases that enter the insect respiratory system and kill all life stages by disrupting cellular processes (Chapter 8).

We may divide insecticidal sprays into two groups. *Space sprays*, for the control of insects in flight by insecticides dispersed in the atmosphere: these include aerosols, fogs and smokes. *Surface sprays* are for the control of insects resting or crawling on surfaces by insecticides applied directly to a surface. Space and surface sprays may be separated on the basis of their spray droplet size.

Spray type	Dispersion	Droplet size (diameter μm)
Space spray	Fog/smoke	<5
	Aerosol	<50
	Mist	50–100
Surface spray	Mist	50–100
	Fine spray	100–400
	Coarse spray	>400

6.2 FORMULATION

Space sprays are usually formulated as:

- *solutions*, usually in a light oil such as odourless kerosene or
- *emulsions*, where the insecticide is usually dissolved in an oil base which, on mixing with water, distributes as fine droplets.

Space spraying will kill all the insects it contacts. It will also effectively decrease the total population, provided it is repeated at an appropriate frequency, which will depend on the particular active chemical, the ambient temperature, the target insect, and the degree of infestation. Many modern space spray formulations increase flight activity and flight time and thus, enhance the chances of insecticidal contact. This hyperactivation or "flushing" action is possessed by many pyrethroids and pyrethrin formulations.

Surface sprays are usually formulated as:

- *solutions* in an oil base; these are useful where water carriers are to be avoided, such as near electrical equipment;
- *emulsions* as described in section 6.2;
- *wettable powders*; where the insecticide is absorbed onto an inert carrier, such as china clay. Surfactants are included in the formulation and enable the powder to be suspended in water. Inevitably, the wettable powder formulation will leave a visible deposit, but this is slight in the case of highly active insecticides, such as pyrethroids, where low active ingredient dose and, correspondingly, lower, powder-carrier dose is needed. Where the powder deposit is likely to cause offence, for example, high-quality polished surfaces, the emulsion is usually employed;
- *suspension concentrates*, prepared using solid active ingredients, most are liquids or semi-liquids. The active ingredient is ground with a finely divided solid carrier and suspended in a liquid phase such as water;
- *microencapsulation*, where the insecticide is protected by coating it with a material such as gelatine. This breaks down on the target surface to release the insecticide;
- *dusts*, utilizing an inert carrier, such as natural diatomaceous earth or synthetic silica, as the diluent for an insecticide which is present at a low concentration in the ready-to-use product.

In all formulations the *active ingredient* (ai) is the insecticide. The other chemicals used, such as surfactants, emulsifiers, solubilizers, synergists, and diluents, are known as the *inert ingredients*. Inert is often an erroneous term since these chemicals may be chemically and biologically active, as is silica, for example.[269]

6.3 ACTIVE INGREDIENTS

Many insecticides have been evaluated in the laboratory against *L. serricorne*. These include formulations with the following active ingredients: chlordane,[303] dichlorvos, trichlorfon, chlorpyrifos, pirimiphos methyl, pyrethrin, nicotine, resmethrin, permethrin, cypermethrin,[47] fenvalerate, biphenrin,[49,208] tetrachlorvinphos,[127,129] tralomethrin,[44,48] bromocyclen.[127] Similarly, laboratory studies have evaluated formulation performance[48] and the effect of temperature on insecticide toxicity.[49,331]

Tobacco seeds have been protected using malathion,[268] and unconventional insecticides show promise for use on-farm, such as bacteria like *Bacillus thuringiensis* for *E. elutella*[50,64,282] and virus.[194] Indeed, bacteria show promise for future investigation.[188]

Dichlorvos (DDVP) was widely used in warehouses[44,63,160,222,244,276] against *L. serricorne* and *E. elutella*, but reliance is now on sanitation, fumigation, methoprene, and space sprays.

Where an insecticide is required, the ai of choice is natural pyrethrins or a synthetic pyrethroid. Some of the earliest work to develop space sprays used pyrethrum in mineral oil to control stored product moths[274] dosed at 17 mg ai/m^3. Pyrethrum mists and aerosols were applied at dose rates ranging from 5–35 mg ai/m^3 for the control of *E. elutella* and *L. serricorne*.[239,338]

In 1946 piperonyl butoxide was shown to synergize the insecticidal properties of pyrethrum. Although not an insecticide itself, piperonyl butoxide blocks the processes by which insects render insecticides ineffective by oxidative metabolism, thereby increasing their potency. Thus a 1:10 pyrethrins: piperonyl butoxide spray is equivalent, against stored product beetles, to 4.4 × pyrethrins alone.[67,126] This allows effective dose rates of 1 mg ai/m^3 for *E. elutella* and 4.8 mg ai/m^3 for *L. serricorne* [159] which significantly reduces the volume of ai sprayed.

Use of the thermal fogging technique[72] utilizing high-volume rates of application has declined over recent years, partly due to the risk of fire from the combustion chamber of the fogger, and the larger than desired droplet size. Ultra-low volume (ULV) or controlled droplet application (CDA) aerosol generation [338] is the preferred means of application as droplet size may be controlled (section 6.4). Application rates of 0.2 ml of formulation/m^3 are employed. Due to low concentrations of the ai in the spray formulations available in some countries, high application rates may still be necessary. Under all circumstances application rates given on the product data sheets/labels **must be followed**.

Today, the use of natural pyrethrins in space sprays has been replaced in some areas by the use of synthetic pyrethroids, such as permethrin [70] and deltamethrin. In some countries, such as Germany, the use of pyrethroids is not permitted for stored products. Indeed, with a properly managed infestation control programme, space sprays may not be necessary. It may be felt that space spraying hides the problem by decreasing pheromone trap catches by killing flying adults before they may be caught and leaving the larval infestation untouched. However, if larval infestation is present, emerging adults will soon appear on traps and so betray its presence. An example of the role insecticides may play is given in section 9.5. Similarly, surface sprays become obsolete as sanitation standards improve since "where one may spray, one may clean!"

Only insecticides labelled for use against *L. serricorne* and/or *E. elutella* may be considered when allowed for use in warehouses and/or factories. Often a blanket approval for crawling and flying insects in food processing facilities will be given. Insecticides available in some countries are formulations including: methoprene (Dianex ®-Chapter 7); pyrethrins (Pybuthrin®); pyrethroids (K-Othrin®; Reslin®). Consult your local or nationwide pest control association, government agencies or current contractor to find out the availability of such insecticides.

6.4 APPLICATION EQUIPMENT

Space sprays may be applied as:

- *smoke* or pyrotechnic mixture which vapourizes the active ingredient, which then condenses to form an insecticidal smoke;
- *fog* from thermal fogging machines: these generate large quantities of hot air into which is injected the insecticide formulation. This vapourizes the formulation which then condenses in the cooler ambient air to form the fog. Oil solutions produce dense fogs, water-based formulations, a haze;
- *aerosol* by generators that use a high volume of fast-moving air to fragment the insecticide formulation into an aerosol. Aerosols may also be generated using high-pressure carbon dioxide cylinders or aerosol-pressurized canisters for small-scale applications. Because the droplets are small, good distribution is attained. This has led to the use of lower volumes of insecticide at a higher concentration. This high concentration to low volume relationship is the basis of the ULV spraying technique. The most important aspect is to have controlled, uniform droplet size to obtain maximum insecticidal efficiency via CDA.[238]

Surface sprays may be applied via:

- *knapsack machines*, with hydraulic pumps that move insecticide from the reservoir tank to a nozzle for dispersal. These are operated by hand as spraying progresses. Motorized sprayers utilize an electric or internal combustion power source to operate a pump for large-scale spraying, such as those used on farms;
- *compression sprayers* or pneumatic sprayers are hand-held sprayers which incorporate an air pump to pressurize the reservoir before spraying commences;
- *mist blowers*, motorized units which use an air blast to atomize the spray into a mist and propel it to the point of impact on the target surface. Such sprayers may be capable of propelling the spray vertically for 10 m;
- *dusters*, which may be spatulas for sprinkling, "pepper-pots", hand-

operated mechanical dusters, such as bellows dusters or, for large-scale applications, motorized mist-blowers converted for dusting.

6.5 APPLICATION TECHNIQUES

Insecticides must be applied uniformly at the correct dose or application rate, either in the air space or on the surfaces being treated. Generally, space sprays may be considered when product is at risk from insects entering the factory or warehouse and to control adults which have emerged from bales etc. However, screens and curtains on building openings will effectively prevent entry.

Surface sprays may be considered for areas that are impossible to clean, such as wall voids, until structural modification allows adequate sanitation.

Insecticide treatments require a thorough understanding of the insect's life cycle and its distribution in the infested premises. Monitoring techniques exist to assess *E. elutella* and *L. serricorne* infestations and direct corrective action (Chapter 3).

It cannot be over-stressed that it is important to inspect and monitor insect populations in order to determine their presence and pinpoint areas of infestation. Further visual inspection with a torch within areas of the building, or within tobacco cases, may then be made and appropriate measures taken.

6.5.1 INFORMATION AND TRAINING

Before commencing any spraying operation, all personnel involved in the treatment must be trained to handle the insecticide correctly. Where necessary, the applicators must have appropriate licences to apply insecticides. Specific product information can be obtained from labels, instruction leaflets, materials safety data sheet (safety information sheet), and manufacturers' technical information, etc.

These documents are obtainable from the manufacturer or supplier and provide full details of how the product is handled and applied, including guidance on personal protective clothing and equipment requirements. Details should address how to deal with any incident which may arise, such as fires or spillage, how to dispose of the product, and essential first aid information.

6.5.2 SPACE SPRAYING

Planning

Each insecticide will differ as to the quantity required per unit volume. The space spray[117] dose rate is calculated for the total volume of the building to be treated. For example if the building volume is 50,000 m^3 and the dose rate is 200 ml/1,000 m^3, we will need 10 litres of ready-to-use spray.

The spray time, based upon an application rate of 0.1 l/minute, will be 100 minutes. The application rate will depend on the ULV aerosol generator settings. Thus, for any insect species and item of application equipment in a specified building, the time required may be defined.

Treatment

The activity of space sprays depends on direct contact of the spray droplets with the flying insects, so treatments should coincide with the maximum flight activity of the insects, usually after dusk. Pyrethroid sprays exhibiting "flushing" activity will stimulate insects and increase flight activity, thus improving their performance. Space sprays may have a considerable effect on crawling insects in harbourages. Space spray labels generally recommend particular attention is paid to insect harbourages, thus increasing the efficiency of control. The space spray has a limited capacity to penetrate harbourages and will not penetrate hogsheads, bales, cases, cardboard boxes or dust accumulations in hidden spaces such as in or beneath machinery. This is the reason why space sprays can be regarded only as an aid in the control of crawling insects. Space sprays will not control the adult or immature stages of insects living within packaged tobacco, such as *L. serricorne* and *E. elutella*.

Inevitably, insecticides applied by space spraying will settle on surfaces. However, the relatively small amount of chemical deposited will rarely reach insecticidal levels. All tobacco, stocks of product, materials and processing surfaces, such as conveyor belts, must be covered to avoid possible contamination.

Areas to be treated must be vacated by all personnel before spraying, except those engaged in the treatment. Doors and windows must be closed and, where appropriate, air conditioning or circulation shut off. Spraying should commence at a point furthest from an identified exit and gradually move towards the exit as the treatment progresses. When using electrical spraying equipment lay out the lead to its full extent and move the machine towards the power point as the spraying progresses. The aim is for the operator to remain in untreated air, always leaving the treated area behind. Apply the spray evenly throughout the area. Treated areas must remain closed for at least 30 minutes following the completion of spraying (or for the period directed on the product label). Ventilate the remaining spray before allowing the area to be re-occupied.

The lack of residual insecticidal activity following space spraying means that treatments should be repeated as frequently as the situation demands. The need for further treatments can be determined from the results of monitoring activities' data. Local laws may require the above sequence to be modified. Users must ensure that the insecticides they use comply with local regulatory requirements and are applied according to the appropriate product instructions.

6.5.3 SURFACE SPRAYING

Planning

The surface spray dose rate is calculated for the area to be treated. Typically, an application rate for the ready-to-use spray of 5 l/100 m^2 is used. When treating impervious or highly porous surfaces, however, much less or more spray might be required to achieve an acceptable treatment on these surfaces. By applying surface sprays to a point just short of run off, account is taken of the different absorbency of surfaces noted earlier. This is a practical interpretation of an application rate which is otherwise difficult to judge at the operational level where surfaces are not encountered in easily measured units, i.e. base of machines and/or roof support pillars. This application method is readily accepted by regulatory authorities. Operators must be trained in the correct speed of application to deposit the needed amount of insecticide.

The actual dose rate is equal to the concentration of the ai in the spray liquid times the application rate. Thus, a dose rate of 100 mg ai/m^2 at an application rate of 5 l/100 m^2, requires the concentration of the spray liquid to be 2 g/l (0.2%). It is difficult to quantify the application of insecticidal dusts, although values ranging from 0.5–30 g/m^2 may be given on product labels. In practice, dusts are sprinkled or blown on to surfaces or into harbourages.

L. serricorne is more difficult to control than *E. elutella*. This is reflected in the dose rate recommendations and must be considered when planning a pest control strategy. The need for a more diligent treatment at the commencement of a spraying regime may demand that a higher dose rate be used initially. A lower "routine" dose rate may be possible once the pest is brought under control. Both are acceptable only if allowed by the label.

Surface sprays are usually applied to a surface in anticipation of the pests crawling over the surface and picking up a lethal dose of insecticide (indirect contact action). This requires a residual level of insecticidal activity that depends upon a number of factors, including:

- *active ingredient* – some stay active longer than others;
- *formulation* – influencing the behaviour of the insecticide on the surface and its ability to be picked up by the insect;
- *efficiency of application* – was the insecticide applied correctly to the appropriate surfaces?

- *nature of target surface* – the porosity and the chemical composition of the target surface will influence the performance of the insecticide. Dry, clean surfaces free of dust and grease are desirable;
- *treatment of target surfaces* – surface sprays are generally only loosely associated with surfaces and are therefore vulnerable to cleaning, flooding, or being covered by dust;
- *insect species* – the varying levels of susceptibility of insects have already been noted. This may be due, in part, to the ease with which the insecticide formulation is picked up by different insects.

It is well known that some surfaces are hostile to the survival of insecticides, especially those which are chemically active or are porous. It is desirable to maintain an undisturbed surface deposit of insecticide for as long as possible so that it is picked up by insects which cross the treated surface. Recent work has indicated that the residual activity of some commonly used insecticides may be severely limited. It has also confirmed the superiority of particulate formulations, such as suspension concentrates and wettable powders, compared with wholly liquid formulations, such as emulsions and solutions.[29,73]

Treatment

Usually a water-based formulation will be used. However, a need to avoid problems, such as staining, corrosion, damage to valuable surfaces or damage to electrical equipment, may lead to the choice of an oil-based spray or dust. A compromise may be inevitable. Dusts are particularly useful where access is limited and they can simply be blown into the area needing treatment. Two formulations, chosen for their particular strengths, may be used to ensure a thorough treatment (for example, a water-based spray in accessible areas and a dust in the inaccessible voids). Keep personnel **not** involved in application of the insecticide out of the area during the treatment and until the residues are dry. They must not be contaminated with the insecticide during the treatment or allowed to contact the deposits while they are still wet. Similarly, raw and manufactured products must not be contaminated with the insecticide. Further treatments may be necessary based upon the results of surveys to monitor the insect population.

Insect growth regulators 7

Methoprene is an insecticide or insect growth regulator (IGR) possessing the ability to halt the life cycle of target insects. The compound mimics the effects of the insect's juvenile hormone, hence it may also be termed a *juvenoid.* IGRs do not eliminate the insect in the manner of conventional insecticides (Chapter 6). The mode of action is to prevent the larval form of the insect from becoming an adult.

Some individuals identify methoprene as a "miracle" product and expect it to be the ultimate agent in solving infestation problems in one easy process, and quite possibly, even dispense with other control measures. This is not the case: methoprene products, with their positive attributes, are only additional tools in aiding the tobacco industry in the task of protecting their raw material and providing the customer with a higher quality product. Other IGRs are likely to find use in the tobacco industry: at this time, however, methoprene is the only one in widescale use.

7.1 FORMULATIONS OF METHOPRENE

Methoprene posseses a sesquiterpenoid carbon skeleton and is composed solely of carbon, hydrogen and oxygen. The formula is:

Isopropyl (2*E*, 4*E*)-11-methoxy-3,7,11-trimethyl-2,4-dodecadienoate

Methoprene was registered by the USA's Environmental Protection Agency in March 1975 as Altosid SR-10®, to control flood-water mosquitoes. This was the first registration of an IGR anywhere in the world. Since then, many products containing methoprene have been formulated and proved efficacious in controlling their respective target insects.

Two formulations, Kabat Tobacco Protector and Dianex Residual Spray, are used to control *L. serricorne* and *E. elutella.*

7.2 CHEMICAL AND PHYSICAL PROPERTIES OF KABAT[375]

Kabat IGR is a clear, blue–green, liquid with the characteristic odour of alcohol.

Specific gravity: 0.814; Molecular weight: 310.48; pH (1:10 dilution in water): 5.2;

Vapour pressure: 40 mmHg at 19°C; Boiling point: 78.5°C;

Flash point, undiluted formulation: 15.6°C; Flash point, dilute spray mix: 44.4°C;

Viscosity (Brookfield LB, #1 sindle, 20°C, 30 rpm): 3 cps;

Miscibility: water <25% (v/v): soluble, >25% (v/v): insoluble;

Most organic solvents: completely miscible;

Percent ai methoprene: 5.0%; percent inert ingredients: 95.0%, ethyl alcohol;

Weight of ai per litre: 41 g;

Corrosion: rated excellent; negligible in Monel 400, copper, yellow brass, mild steel, stainless steel 304 and 316;

Storage stability, undiluted formulation: stable from room temperature to 42°C for 1 year; stability, dilute spray mix: short-term, requiring use within 48 hours.

7.3 METHODS OF CONTROL

Kabat Tobacco Protector (Kabat) is used in the tobacco industry to protect the treated tobacco during storage and prevent *L. serricorne* and *E. elutella* population increase. Adult insects may colonize methoprene-treated tobacco and lay eggs. Larvae hatch from the eggs and begin feeding on the tobacco, usually burrowing beneath the surface of infested bales. The larvae pass through several phases of growth (instar), the duration of which depends upon ambient temperature and humidity (Chapter 2), before entering the pupal stage. Under normal conditions the larvae within the pupal cell metamorphose into the adult form. However, larvae which have ingested methoprene-treated tobacco during the fourth instar[231] fail to emerge as adults, instead forming quasi adults or adultoids, and successive generations do not develop. Hence, the compound interrupts the developmental pathway and breaks the life cycle. Some evidence also exists that methoprene may have an ovicidal effect[231] and egg fertility effect.[237]

Methoprene was first evaluated in the laboratory, to control *L. serricorne*, in 1973, and was formulated as Altosid-treated dog food at 5 ppm methoprene.[329] Since then, efficacy has been confirmed by others.[2,14,74,235,244] Indeed, 10 ppm of methoprene will protect the tobacco and prevent infestations of *L. serricorne* and *E. elutella* for up to 4 years.[234,235,309] Lower doses of 5 ppm were later demonstrated to be effective for more than 2 years[232] and this is recognized as a minimum target for application.[296] Some laboratory studies indicated much lower doses to be effective;[45,46] however, the transition from laboratory to the field or stemmery requires a higher dose due to imperfect distribution of the methoprene. The current industry standard allows for tobacco treatment at 5 to 10 ppm which is achievable with modern application equipment.

7.4 TOXICITY AND DEGRADATION

Methoprene acts specifically against insects and is virtually non-toxic to vertebrates. The LD_{50} for methoprene is 34,600 mg/kg: the largest volume of material that could be administered to the test animal. The World Health Organization has determined that the safety level of methoprene is such that approval was given for its use in drinking water at 1 ppm for mosquito control.

The fate of methoprene in the environment is short, a half-life of 2 days in field water[304,305] and less than 2 weeks when exposed to sunlight photodecomposition.[278] Application to tobacco at low moisture and storage in boxes, ensures a long, residual life.[325] Degradative losses of methoprene occur primarily in the first few months after application, when up to 10% may be lost due to hydrolysis and oxidation. Long term losses are estimated at 30% total over a 2-year period, due to distillation, oxidation and UV degradation. This should be taken into account when deciding dose rates. The distribution of methoprene in cigarette smoke has been studied.[151] Insects metabolize methoprene to its methoxy and hydroxy derivatives by the action of esterase and oxidase enzymes.[123]

7.5 EFFECTS ON TOBACCO QUALITY

The quality of tobacco, when treated with methoprene, is preserved during the storage period and cigarettes manufactured from methoprene-treated tobacco do not reveal off-tastes (Table 7.1). Panelists who participated in taste tests of cigarettes made from methoprene-treated tobacco (10 ppm methoprene) did not find any significant difference in 13 of 15 tests conducted during 1976–1980.[376]

Further taste tests in 1988 confirmed that smokers do not detect any difference in the taste of cigarettes made from methoprene-treated tobacco. Numerous industry tests (unpublished) confirm that methoprene-treated and non-treated tobacco taste the same.

Although not yet approved for use world-wide, methoprene is registered for use in most tobacco producing countries around the world (Table 7.2).

7.6 APPLICATION

Kabat is available in a formulation which contains 41 g/l of methoprene as 5% w/w with ethyl alcohol 95% w/w. Kabat is mixed with water and applied directly to the tobacco during processing at the stemmery. A proportional dilution applicator is calibrated to the tobacco flow-rate through the dryer. The recommended label dosage rate is 200 g Kabat (which contains 10 g of methoprene) per 1000 kg of tobacco, or 10 ppm methoprene. Only the S-enantiomer is biologically active; thus the recommended rate of 10 ppm methoprene will effectively contain 5 ppm active methoprene.[166,167]

Samples must be analysed to determine the efficiency of the application equipment. It is wise to place aside several of the storage containers as a representative sample of the treated lot or crop year, and sample the concentration of methoprene from these designated "test" containers during the storage period. The type of storage container (box, bale, bag, etc.) and conditions within the storage warehouse may affect the methoprene levels on tobacco. Sample the "test" containers for methoprene concentrations at least once per year.

Complete coverage of the processed commodity is assured by using multi-directional nozzles, and by tumbling the tobacco onto the conveyor prior to its compaction in bales or boxes. Although the tobacco is usually

Table 7.1 Taste testing methoprene-treated tobacco cigarettes[376]

Tobacco Cased	Methoprene Uncased	(ppm)*	Panel membership	Findings vs. control
X		1.7	42	NSD‡
X		3.0	31	NSD
X		7.0	46	NSD
	X	17.0	17	NSD
	X	20.0	17	NSD
X		8.0	17	NSD
X		5.0	17	NSD
X		2.6	17	NSD
X		2.6	14	NSD
X		3.0	13	D
X		3.7	12	D
X		3.0	14	NSD
X		3.7	14	NSD
X		3.0	445	NSD
X		3.7	419	NSD

* Methoprene in cigarette
‡ NSD = No significant difference; D = Difference detected.

Table 7.2 Countries where Kabat and Dianex are registered and or legally available

Region	Countries
Kabat	
Africa	Kenya, Malawi, Nigeria, Sierra Leone, South Africa, Tanzania, Uganda, Zimbabwe
Europe	France, Greece, Italy, Switzerland, Turkey
Americas	Argentina, Brazil, Canada, Chile, Colombia, Dominican Republic, Ecuador, Guatemala, Honduras, Mexico, Peru, Uruguay, USA, Venezuela
Australasia	Australia, China, Hong Kong, Indonesia, Malaysia, Thailand.
Dianex	Holland, Japan, South Africa, Switzerland, USA.

treated as it leaves the dryer, other alternative sites are being explored. These include oriental tobacco treatment pre-packing and big-leaf treatment before the drier. Treatment of the oriental bale surface gave only partial control.[134]

7.6.1 SAFETY DURING HANDLING

Since Kabat is 95% alcohol, the formulation is flammable and should be handled as such. Product labelling recommends that personnel handling Kabat containers, prior to mixing with water, wear gloves and eye protection. After mixing with water for application at a ratio of approximately 1:25 there is little danger from flammability.

7.6.2 APPLICATION EQUIPMENT

The original application equipment, a model K-104, has been replaced by more precise and sophisticated equipment, available from W.H. Dickinson Engineering Limited, GBE LEGG Ltd. and Chemtrade International Ltd.

The Chemtrade unit is designed for use where there is low, flow-through volume of the tobacco, or in application prior to a blending or bulker silo. This type of application equipment is also advantageous in applying Kabat to Oriental tobacco where there is no re-ordering or dryer chamber, but only a blending box. Other units have been designed and built by engineers in individual tobacco processing companies.

Whatever the application equipment, methoprene must be applied uniformly to ensure that each piece of tobacco has been treated with the target application rate. This "exact" application requires special consideration to the design, operation and maintenance of the application equipment. The equipment and procedures described here have been proven to achieve consistent and uniform application rates at target levels.

A process flow diagram of a typical Kabat application system is shown in Fig. 7.1. Each component in the system is necessary to ensure accurate metering of methoprene onto the tobacco.

7.7 RESIDUE TESTING

Tobacco samples must be taken during application and storage for determination of methoprene residues on the tobacco. Take core samples vertically downwards through the tobacco bale in order to obtain a value for the average application rate. Grab samples taken sideways through the bale will indicate the variation of the residue within the tobacco package. The vapour pressure of Kabat is such that only limited migration may occur within the bale.

Accurate analytical methods based on gas chromatography and/or high performance liquid chromatography (HPLC) are expensive, require trained laboratory personnel, and are sometimes located thousands of miles from the processing plant.[163] CORESTA has standardized the chromatographic methods.[108] With these procedures, however, there is a long delay between methoprene application and obtaining results for calibration of application equipment. Other, more rapid methods of determining methoprene residues on site are under development. These include rapid HPLC[372] and enzyme-linked immunoabsorbent assay (ELISA) kits.[164]

7.8 STORAGE

Tobacco with sufficient methoprene does not require further insect control. However, methoprene is light sensitive and treated tobacco must be protected from the light.

Store treated tobacco separately from untreated tobacco to prevent cross-infestation within the warehouse or storage complex. Since *L. serricorne* has been documented to fly distances of 3 km, isolation of the methoprene-treated tobacco is necessary. If isolation of treated tobacco is not possible immediately, other tobacco in the same location (warehouse or storage complex) must be fumigated. Methoprene-treated tobacco which is kept separately is available for immediate use in production. There is no down time for further treatment, as is the case when warehouses must be closed for fumigation; neither is there an inconvenience to personnel in the storage area.

7.9 RESISTANCE

Insects will eventually develop resistance to any pesticide, or resistance will be selected within the insect population.[310] **It is imperative to apply methoprene correctly and minimize the possibility of selecting for resistance.**

7.10 DIANEX®

While methoprene is used primarily for direct application to tobacco being processed in the stemmery, it is important as a premise application to control *L. serricorne* in tobacco processing manufacturing facilities. This is a supplement to, not an alternative for, good sanitation (Chapter 4).

An 89% reduction in insect-related consumer complaints was reported after initiating the use of methoprene as Dianex in space spraying the factory atmosphere (Fig.7.2).[176] In these studies Dianex was mixed with water (1 ml/90 m^2 of surface space) and fogged in the factory atmosphere

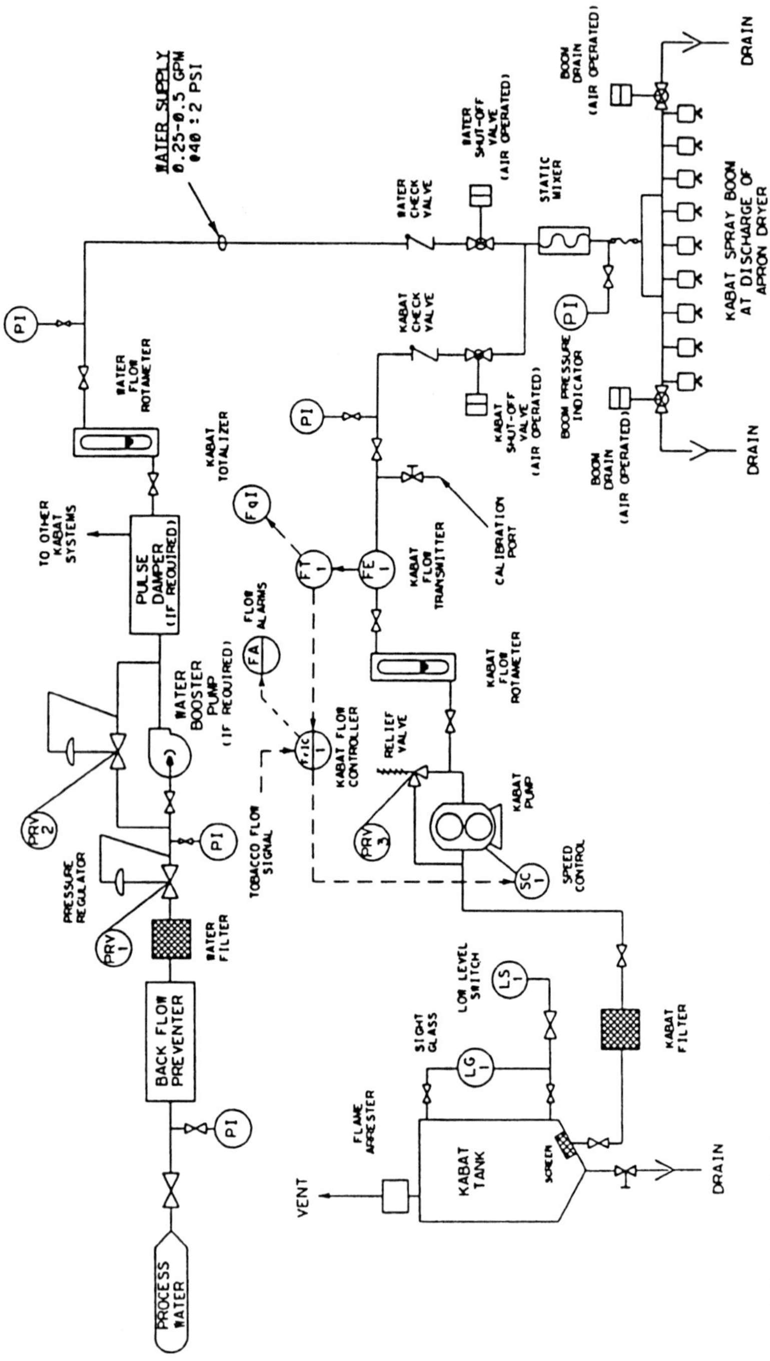

Fig. 7.1 Kabat application equipment component flow diagram.

KABAT application equipment components

Blackflow preventer stops non-potable water from entering the potable water supply system. This is normally a plumbing regulation. The use of reduced pressure backflow preventers is preferred to an air gap tank to eliminate the possible microbiological problems associated with the tank.

Boom drain valve gives anti-drip protection at the nozzles by relieving boom pressure during shutdown conditions. It may also be used for flush cleaning the boom with water.

Boom pressure indicator indicates liquid pressure in the spray boom. Correct pressure is important to ensure proper atomization at the spray nozzle.

Calibration port allows the calibration of the Kabat flow transmitter.

Flow alarms alerts the operator to conditions when Kabat is over or under applied.

Kabat check valve prevents the back flow of water into the Kabat system.

Kabat filter prevents particulate matter from getting into the Kabat pump and the spray nozzles.

Kabat flow controller allows ratio control of Kabat to the tobacco flow input signal from the weighbelt. Based on the Kabat target application rate (ppm), the controller calculates the proper Kabat flow and controls the speed of the Kabat pump to achieve the desired flow.

Kabat flow rotameter gives a coarse measure of Kabat flow for monitoring process conditions.

Kabat flow transmitter gives a precise measure of Kabat flow for accurate flow metering and quantification of total amount of Kabat used.

Kabat pump supplies low volume Kabat flow at sufficient head pressure to achieve accurate metering.

Kabat shut-off valve automatically shuts-off Kabat during process shutdown and upset conditions, and also prevents the leakage of Kabat back into the water system.

Kabat tank is a closed tank that holds Kabat for application. Equipped with explosion-proof vent and bottom filter for coarse screening.

Kabat totalizer indicates the total amount of Kabat used up to the current time.

Low level switch prevents air from entering Kabat metering system by shutting off the Kabat pump before the tank is completely empty.

Pressure relief valve prevents over pressure of the Kabat flow system when the Kabat shut-off valve is closed.

Sight glass on the Kabat tank indicates the level of Kabat in the tank. Used by operator when filling the tank.

Speed control shows the precise adjustment of Kabat pump speed to achieve accurate Kabat flow metering.

Static mixer gives uniform mixing of Kabat and water just before application.

Water booster pump increases water pressure to the level required for proper atomization at the nozzles. This pump is required only if the process water supply cannot maintain a pressure above 6.5 kPa.

Water check valve prevents back flow of water and Kabat into the water system.

Water filter removes particulate matter from the water that can cause nozzle clogging in the spray boom.

Water flow rotameter gives a coarse measurement of water flow for monitoring process conditions.

Water pressure regulator regulates the water supply pressure to minimize pressure changes that affect spray nozzle atomization.

Water pulse damper minimizes pressure fluctuations in the water system when more than one Kabat system uses the common water supply. This device is required when two or more Kabat application systems, or other factory devices, are sharing a common water supply.

Water shut-off valve automatically shuts-off of the water during process shutdown and upset conditions.

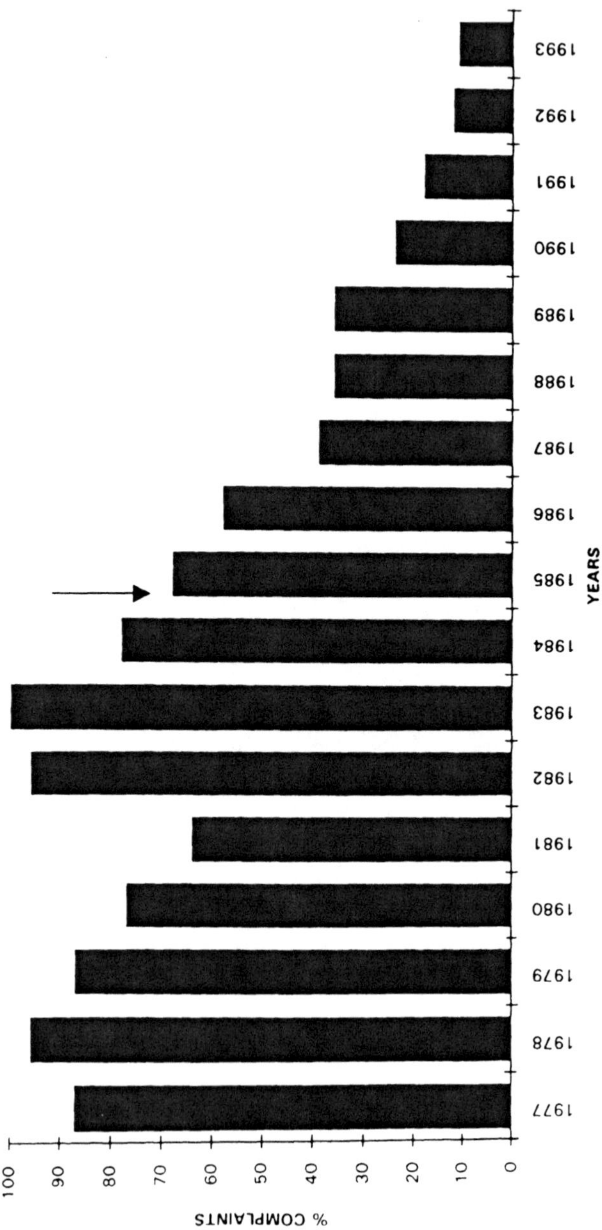

Fig. 7.2 Percentage of insect-related consumer complaints showing the initiation of Dianex fogging (arrow) in 1985.

using a London Aire ULV machine. This treatment was scheduled monthly to maintain 10 ppm methoprene concentration on surface dust. Dianex is also commonly applied using pressure knapsack sprayers or directional ULV-type equipment (see Chapter 6). In some countries Dianex is substituted by Kabat to which is added suitable emulsifiers.

Fumigation 8

L. serricorne or *E. elutella* infesting tobacco are killed by temperature treatments (Chapter 5) or fumigation. Fumigants are gases which penetrate the tobacco mass and, when used correctly, kill 100% of all life-stages of the insects within. Only two true fumigants are used: phosphine, and to a lesser and declining extent, methyl bromide,[82] although, other fumigants have been used.[83] Fumigation by controlling the basic atmospheric components is used on a small scale. Therefore, we will deal with phosphine in detail and then show how it differs from the other fumigants.

These agents will kill most life-forms, including humans; therefore, they are regulated by law. Such laws and maximum safety procedures must be followed. Fumigation requires trained, experienced, specialist personnel. What follows is designed to help tobacco industry personnel interact knowledgeably with fumigators.[15]

Attempt fumigation only with comprehensive training.

8.1 PHOSPHINE

Phosphine (PH_3) or hydrogen phosphide has the following physical properties: molecular weight, 34.0; boiling point, –87.7°C; melting point, –133°C; flashpoint, 100.0°C; low water solubility; specific gravity (air= 1.0), 1.214; explosion point in air, 1.79%.[52] It is a colourless gas that occurs in the hydrospheric phosphorus cycle[87,114] and was discovered in 1785.[145] Phosphine is toxic to animals and kills by inducing lipid peroxidation with free radical involvement.[75] While all insect life-stages are killed by phosphine, the egg is usually the most tolerant.[172,356] Toxicity is related to temperature.[175]

Phosphine fumigation of tobacco began in Europe in 1959[355] followed by corroborative trials in Japan and the USA. Large-scale use by the worldwide tobacco industry was established by 1975 and continues today.[78,81,85,89,156,199,244,248,335,347] The correct use of phosphine is essential if we are to avoid the selection of resistant strains[337] and eliminate any occurrence of tolerance.[374] Resistance in beetles is due to the active exclusion of, or the ability of insects to eliminate, phosphine.[277] **To date there has been no record of a proper fumigation failure due to resistance.**

Various formulations based on aluminium phosphide[150] and magnesium phosphide[253] are available. Phosphine gas is evolved following exposure to atmospheric moisture:

Aluminium phosphide	+ moisture	⇒	aluminium hydroxide	+ phosphine
AlP	+ $3H_2O$	⇒	$Al(OH)_3$	+ PH_3
and				
Magnesium phosphide	+ moisture	⇒	magnesium hydroxide	+ phosphine
Mg_3P_2	+ $6H_2O$	⇒	$3\,Mg(OH)_2$	+ $2PH_3$

Aluminium phosphide was used exclusively until the introduction of magnesium phosphide in the early 1980s.[288,364] Phosphine gas evolution is dependent on temperature and humidity. Evolution is more rapid from magnesium phosphide. Aluminium phosphide formulations also contain ammonium carbamate which liberates ammonia and carbon dioxide which helps burst the pellet and tablet formulations. All formulations release a characteristic odour of carbide or garlic, detected by the nose at 0.02 ppm, long before phosphine gas reaches 0.1 ppm level. When fumigating correctly one will not be able to smell it.

The formulation of choice should allow ease-of-handling to speed application and removal of residues. Packaged formulations (i.e. bags and plates) allow rapid placement and minimize potential tobacco contamination by containing the spent fumigant residues in the bag or plate (section 9.3).

Magnesium phosphide plates react more completely, leave smaller unspent residues after fumigation, and may be deactivated more completely and easily than aluminium phosphide bags. Magnesium phosphide plates are preferred for the above reasons.[131,253,288,332] When these are not available, aluminium phosphide bags may be considered allowing that the fumigation duration will be extended, and deactivation of unspent residues may require weeks and/or considerable labour to puncture the bags to facilitate water entry. The rapid evolution of phosphine from magnesium phosphide fumigants is sometimes given as a disadvantage when fumigating known leaky structures. Such structures should be sealed adequately to allow proper fumigation.

Numerous trademarked formulations are available world-wide. However, the packaged magnesium phosphide formulation most widely used today is the Detia Degesch plate or Fumi-Cel Plate® which are available singly or 20 joined together to form the strip or Fumi-Strip®, to be used complete and not cut. Aluminium phosphide bags are, however, available from a number of manufacturers.

The usual dose of phosphine, 1 g/m^3 (700 ppm), is supported by laboratory and field trials.[35,89,90,91,131,189,229] Here, Haber's rule (see section 8.4) does not apply, as it does for methyl bromide[40] therefore, the fumigation duration should not be shortened.[43] It is **essential** to monitor the gas concentration and ensure that 200 ppm phosphine is maintained for at least 48 consecutive hours, within the greater than 4-day fumigation, to ensure 100% insect kill.[91,172,175] Magnesium phosphide plates will be spent after a 96-hour exposure. When using aluminium phosphide bags the duration of the fumigation to release all the phosphine and achieve 200 ppm for 48 hours will be longer at lower temperatures. For example at between 7°C and 15°C a 10–15-day exposure is necessary.

Fumigations may be conducted when the ambient temperature is above 7°C.

Phosphine has no effect on the flavour, colour or taste of either cased or uncased tobacco, nor does it leave residues.[349]

Phosphine may react with certain metals and cause corrosion,[53] especially at higher temperatures and relative humidities. Metal such as copper, brass and other copper alloys, and precious metals such as gold and silver are susceptible to corrosion by phosphine.

Small electric motors, smoke detectors, brass sprinkler heads, batteries and battery chargers, forklifts, temperature monitoring systems, switching gears, communication devices, computers, calculators and other electrical equipment may be damaged by this gas. Phosphine will also react with certain metallic salts and, therefore, do not expose sensitive items such as photographic film, some inorganic pigments, etc. Aluminium and nickel do not appear to be affected by exposure to phosphine, and zinc reacts only above 90% RH.

All electrical installations within a structure to be fumigated with phosphine must be either protected by sealing or removed from the premises prior to exposure. Water sprinkler heads of fire control systems must be either sprayed with an antioxidant or covered with a wax (paraffin) to keep them from becoming corroded. Sprinkler heads made of nickel, stainless steel or chrome-plate are available. If in doubt, conduct a pre-test. The oxides that are formed on exposure to phosphine can be removed by washing the affected parts with a solution of ammonia in water. It should be noted that the pits which form on the oxidized metals are not affected by the ammonia treatment.

Cigarette stocks suspected of being infested or infested returned goods, not for resale, may be fumigated with phosphine. The procedure is identical to that for fumigating stored tobacco. However, cigarettes with charcoal filters should not be exposed to phosphine. The charcoal will absorb and retain the fumigant, and in some instances lower its concentration to an ineffective level.

8.2 GENERAL TECHNIQUES

Fumigations are carried out in a closed space (fumigation site) which is at atmospheric pressure at all times (see specific gases for possible exceptions). Tobacco fumigations may be conducted in: a warehouse; the volume under a gas-tight (polyethylene) sheet, sometimes called a tent or tarpaulin; a specially constructed chamber; freight containers, trailer bodies or railcars. **The closed space to be fumigated should be sealed to contain the gas – an obvious statement, but one that is often overlooked.**[20] Some tobacco may be wrapped in polyethylene bags (liners), therefore, phosphine will penetrate more slowly. More importantly, the gas may require several weeks to aerate to the required level of less than 0.1 ppm.[80,121,122,137,170,190,213,214,240,366]

8.2.1 SEALING PROCEDURES

In **warehouses** all windows, vents, doors, drains, ventilators, and other cracks/openings must be sealed. Provisions for gas-tight sealing should be included in the design of such closures (e.g. rubber gaskets and clamps). If not, seals must be made each time the building is fumigated. Heavy masking tape at least 5 cm wide, or polyethylene tape (150 microns [μm] minimum thickness) may be applied over closures. Alternatively, sheets of polyethylene may be placed over the entire opening, and secured around the edges with tape. Wrap and seal roof ventilators with plastic sheet that is high density and at least 150 μm thick.

Inspect the building for leaks and potential leaks in floor, walls and roof. Give particular attention to the point at which the roof joins the walls; any openings must be sealed. In the case of double-thickness masonry, a gap is frequently left between the two courses of brick or block. If the top of the wall is not tightly capped, gas may escape through weep holes in the outer course. Sealing may be by taping or by coating with a sealant.

Suitable sealants are asphaltic mastic and "vinylite" plastic (polyvinylchloride/acetate copolymer resin in volatile solvents). Various other commercial sealing compounds may be used, but their gas-tightness and inertness to the fumigant must be established before use. Interior masonry walls must be sealed with at least two coats of vinyl-based paint, epoxy resin, or similarly gas-tight materials.

If it is necessary to fumigate a building to which sealant application is impractical (uncaulked joints in roof or siding, porous masonry, screened louvers, etc.), the whole structure may be wrapped with polyethylene sheeting (at least 150 μm thick). It may only be necessary to wrap sidewalls if the roof is gas-tight. This can be accomplished by using large sheets of polyethylene. Modify such warehouses to allow adequate sealing (above) or use gas-tight tarpaulins within the warehouse.

Sheet, tarpaulin or "tent" fumigation is usually conducted within warehouses. Use a seamless sheet of high-density polyethylene sheet of minimum 150 μm thickness; if in doubt, the material may be tested.[190] If more than one sheet is required to cover the tobacco to be fumigated, tape the seams so as to be air-tight and not pull apart during the exposure period. The seam must be formed from a 30-cm overlap taped at both edges. Sweep clean the floor area involved and anchor the sheet with sand snakes, loose sand or tape. Loose sand, the least convenient material to use, gives the best sealing results, for it causes the sheet to conform with any irregularity in the floor surface. Do not employ solid materials, such as wood, metal, stone, bricks or water snakes, etc. as anchors.

During the draping of the sheet over the tobacco containers, ensure that the sheet is not torn by any sharp corner or edge. Remove, pad, or make smooth all such features. Repair all tears in the sheet with tape. If the tobacco containers are stacked in corners or near walls, make sure that sufficient space exists to permit workers to properly drape the stack and to properly seal the sheet to the floor. Ensure there is no leakage through the floor such as via drains or expansion joints; it may be necessary to lay sheet on the floor (Fig.8.1). Sand snakes must be heavy enough to seal the sheet to the floor.

Introduce gas concentration monitoring tubes into the air space under the sheet or into the tobacco containers by means of a small slit in the sheet, which would then be taped tight ready for the fumigation to begin. Seal the sheet to the floor prior to fumigation with a small section left open to permit placement of the plate or strip. After placement, close or seal the opening. Semi-permanent sheeting is available such as the Bubble® system, from Rentokil.

Fumigation chambers offer advantages including gas-tightness, better control of dosage, and permanency of gas handling equipment. They may be built of metal or masonry and porous materials must be sealed. They may be fitted with exhaust fans. The exhausted gases may be chemically scrubbed.

Freight containers, trailer bodies or railcars may be utilized for fumigation with phosphine.[87,88] New ones should be used, and tape employed for any additional sealing of the body air vents and for the securing of the door gaskets against the door frame of the container.

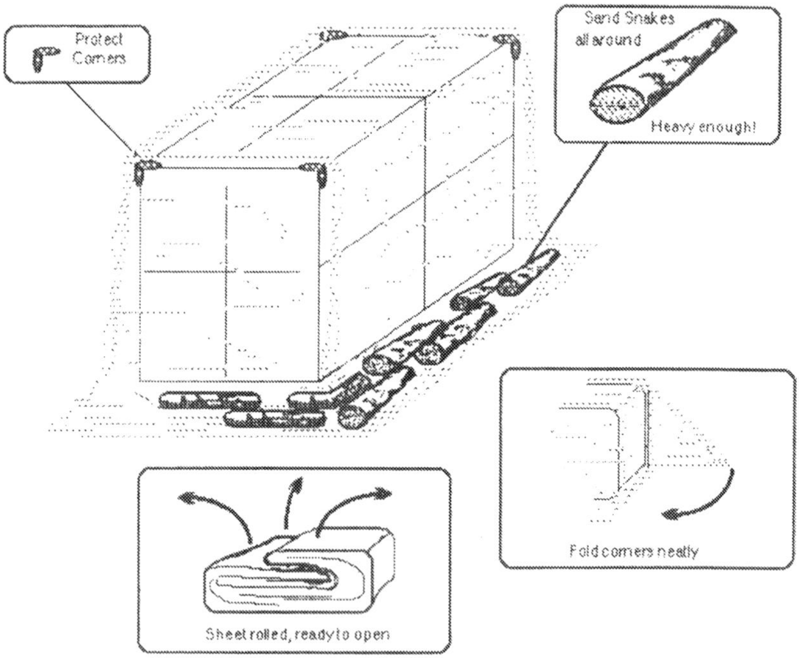

Fig. 8.1 Preparation of a stack fumigation showing the stack of tobacco covered by a sheet which has been folded for easy deployment. Corners are padded to protect the sheet from tears. Sandsnakes overlap and are heavy enough to seal the sheet to the ground. Corner folds are made similar to those when wrapping a parcel. When all stacks are standard sizes the use of sheets becomes simple.

Special attention should be given to making the floor gas-tight. The container should be parked where it cannot be damaged by any moving vehicle or machine, and by at least 20 m from the nearest work site.

Because of the great variation in the gas-tightness of such containers, these fumigations are not recommended. Of course, the whole container may be treated as a tent fumigation as explained previously. Although in-transit fumigations are performed with grain,[112,180] these are not acceptable for tobacco fumigations. While in-transit, phosphine measurements may not be taken, safety risks exist, and problems may occur at destination with unspent residues and containers with danger placards.

8.2.2 SAFETY AND SECURITY

Several rules **must** be followed when using the fumigant.

THE PRODUCT LABEL MUST BE READ CAREFULLY, UNDERSTOOD AND FOLLOWED.

A comprehensive understanding of the fumigant, fumigation techniques and all local laws[11,12] and ordinances must be maintained by the fumigator and obeyed. Trained, licensed, experienced fumigation personnel are the only persons to be considered for such work.

At least **three** full-face canister type gas masks (two for working in pairs and one for rescue) and three self-contained breathing apparatus (SCBA) must be on hand and maintained in good repair. At least two appropriate canisters must be kept on hand at all times for each mask. A record must be kept of the number of minutes a canister is used, and of the concentration to which it is exposed. Minimum standards must be established, such as one gas canister used for 4 hours – respect the product instructions. The canisters must be approved for use with phosphine. Gas masks are used when the exposure to phosphine is likely to exceed 0.1 ppm up to 15.0 ppm. Above 15.0 ppm the SCBAs must be used.[261] Clearly, SCBAs would be used anywhere there is oxygen deficiency, such as a carbon dioxide fumigation (section 8.5).

Equipment for measuring fumigant concentration must be kept at the work site from the beginning of fumigation until the end of aeration. If indicating reagent tubes (section 8.2.4) are used, they must have appropriate ranges for measuring both fumigation levels and trace levels of fumigant. Such tubes have expiration dates printed on them, but if stored in a refrigerator newly manufactured tubes have a shelf life of 2 years.

No fumigation may be performed unless: the supervisor is trained in the use of the specific fumigant, including procedures for first aid treatment of poisoning; the fumigant is legally registered for use in-country; environmental release laws are obeyed; and correct, safe transportation of the fumigant is possible. Protect workers from accidental ingestion:[317] keep the fumigant under "lock and key". Protect workers from potential long-term exposure effects:[153] use respiratory protection whenever phosphine is detectable.

Establish a security perimeter of at least 20 m around all fumigation sites and control entry into this area: only fumigators will be allowed in for the fumigation duration.

At least one guard must be posted for the duration of fumigation and aeration of warehouses, open-air tent and freight-container fumigations. Regular security personnel may be used to periodically check fumigation areas. Such guards must be familiar with all relevant safety procedures for the fumigant. All entrance doors to fumigation areas must be thoroughly posted with warning signs. Signs should be in the local language(s) and include emergency response telephone numbers and the dates the fumigation began and will end.

Local police, fire and health authorities should be notified of all fumigations. A permanent notification may be made in case of regular chamber fumigations. Insurance coverage as required should be reviewed, as some general liability policies may exclude fumigations. All personnel handling any phosphide fumigant will wear dry gloves of cotton or similar material whenever there is the least danger of coming into contact with a fumigant. Contact lenses should not be worn by anyone using a mask. Irreparable damage to the eye may occur if the mask leaks and the gas gets between the lenses and the cornea. Facial hair, such as beards and sideburns, interfere with the proper seating or sealing of a face mask. Such persons should not attempt to use masks or participate in any fumigating activity calling for use of such protective equipment as a matter of procedure.

Never fumigate alone. First aid equipment and procedures must be on site. Inform local hospital, infirmary and medical officers of fumigant type, poison symptoms and medical treatment required.

Open containers of fumigant in the open air. No person is to be allowed in the structure during fumigation except personnel making the application. Bathe and change clothing as soon as possible after having been exposed to a fumigant. Launder all clothing worn during fumigation before being worn again.

All persons likely to come near the fumigation site should be made aware of phosphine properties. The fumigant label and material safety data sheet should be on site at all times. The label may include an additional application manual.

Contingencies must be planned for a rapid response in any emergency. We provide here a number of scenarios with a response. However, those responsible for fumigations should review current practices and develop plans to deal with all possible contingencies including:

Accidental poisoning: a member of the fumigation team inhales gas due to accidental exposure and begins to demonstrate symptoms such as fatigue, nausea, chest pains, etc.
Response: individual will be moved a safe distance from the fumigated area and administered medicinal oxygen if necessary. Individual will be evaluated by a physician, escorted by a member of the fumigation team with a copy of the fumigant label, material safety data sheet and instructions to physician.

Severe weather during fumigation such as wind, rain, hail, and atmospheric inversion.
Response: increase surveillance including additional gas readings and seal evaluations. Do not place or remove fumigant in inclement weather.

Sprinkler system is charged with water or there is an actual flow of water.
Response: contact personnel on the emergency response list. Contact accountable factory/warehouse contracting personnel to respond to fire system problem. Emergency response team will respond to the warehouse (fumigation site) with the following equipment available: fire extinguishers of CO_2 or dry powder type, tongs, gas masks and SCBA, gloves, medicinal oxygen, drums for disposal via wet deactivation, and shovels. Emergency response team enters the warehouse with SCBA to identify the problem. If water is present, plates are to be removed and deactivated. During this activity all persons present must wear SCBA. If any fire is observed, the person on the scene controls the fire and removes the fumigant.

Fire in warehouse
Response: fire will be reported to the fire department, or company fire fighters. A company representative and/or fumigation team member will go to the vicinity of the fire and a guard will be at the main gate to await responding fire fighters. Fire fighters will be taken directly to the warehouse in question. Entry will be delayed until the fire department is on the scene. Respiratory protection will be worn. Plans will already be in place to deal with the emergency release of phosphine, to comply with environmental protection laws.

8.2.3 FUMIGATORS AND CONTRACTORS

Many countries require that fumigations are only conducted by professional fumigators. These fumigators will have passed government-approved examinations and be certified as such. Equally, they will belong to a licensed fumigation company expected to maintain adequate liability insurance. Fumigants can be lethal to man; there are laws restricting their release to the atmosphere. Therefore, it is advisable to use fumigating contractors and to maintain an awareness of your local laws and ensure compliance. Appoint and train someone within your organization to ensure the contractor is fumigating correctly. This applies even in countries where government agents conduct the fumigations.

8.2.4 MONITORING GAS CONCENTRATIONS

Determine gas concentrations within the warehouse airspace and within the tobacco containers during the course of fumigation and aeration. Make gas concentration readings at least at the following hours after initiation of the fumigation:

6, 12, 24, 36, 48, 60, 72, 96 and every further 24 hours until completion (Fig. 8.2).

Place monitoring or sampling lines of 6 mm inner diameter tubing in the corner areas of warehouse space. At least one line should sample the air (minimum of 1 m above floor level) and another line should be attached to a metal tubing extending halfway into the tobacco container or bale. More sampling lines will provide more meaningful data. Extend sample lines to the outside of the warehouse by any available means (holes bored through the wall or door, for example), and cap them with a stopper or tape. Seal the points at which the lines exit the warehouse.

For reading the phosphine concentrations, the lines must be purged of any trapped air by using a suction pump just prior to taking the readings. Any type of suction apparatus, including the gas sampling pump used with the Dräger or Auer tubes, will suffice to purge the lines. A simple calculation of tube:pump volume will indicate the number of pump strokes needed to purge. When many lines are to be purged, consider the use of a motorized suction pump to expedite the process. Purge lines before each successive reading to ensure that the current concentration of gas will be monitored. The number of strokes or interval of time required to purge the lines should be the same as that used in the initial purge. The Dräger or Auer gas detection tubes (or others) should then be employed. Use monitoring tubes such as Dräger, Auer and Kitagawa,[212] although more sophisticated devices such as gas chromatographs and/or electronic devices are available. Tubes are available to monitor 0.01 ppm and above. During fumigations and aerations monitor the airspace around workers and around the fumigation site to ensure no accidental exposures. Other monitoring possibilities exist, such as reactive-paper badges (Dräger), long-term exposure tubes,[308] and electrochemical cells.[120]

Monitor the ventilated airspace until it is less than 0.1 ppm of phosphine. Monitor the tobacco mass until it is less than 0.1 ppm before releasing it for transport/processing. The type of monitoring/fumigation record one would expect to see is shown in Table 8.1.

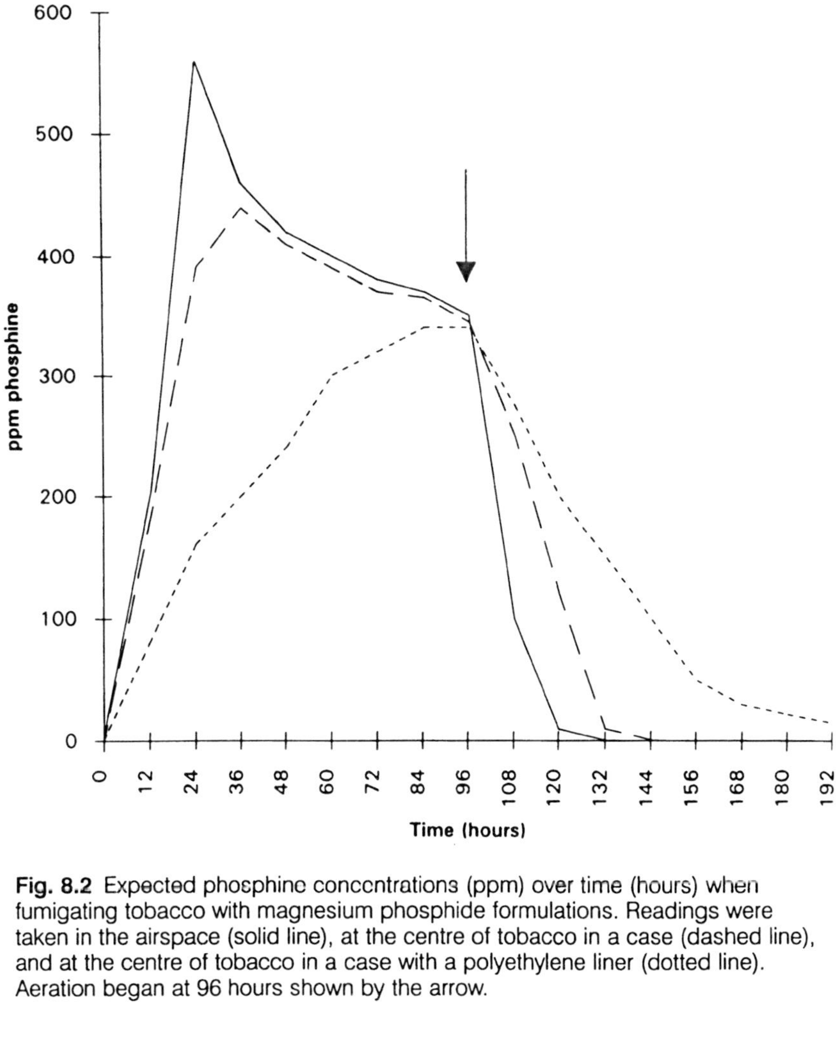

Fig. 8.2 Expected phosphine concentrations (ppm) over time (hours) when fumigating tobacco with magnesium phosphide formulations. Readings were taken in the airspace (solid line), at the centre of tobacco in a case (dashed line), and at the centre of tobacco in a case with a polyethylene liner (dotted line). Aeration began at 96 hours shown by the arrow.

Table 8.1 A typical fumigation record sheet showing the type of information recorded

FUMIGATION CERTIFICATE **NUMBER** __________

	NAME	ADDRESS
TOBACCO SUPPLIER		
CONSIGNEE		
FUMIGATION CONTRACTOR		

TOBACCO DESCRIPTION	
Type:	Net Weight:
Crop Year:	Quantity: [No. of Cases]
Grade:	Poly Lined: Yes☐ No☐
Case Numbers:	

FUMIGATION DESCRIPTION

TYPE: Container ☐ Sheeted Stack ☐ Whole Warehouse ☐ Chamber ☐ Other ____________

FUMIGANT NAME: ____________ FORMULATION: Pellet ☐ Tablet ☐ Bag ☐ Plate ☐ Other ______

TOTAL VOLUME FUMIGATED (m^3):________ DOSE (g/m^3):________ TOTAL FUMIGANT USED (g):______

FUMIGATION DATA [Note: This Section **MUST** Be Completed]

DATES: Fumigant Placed ___/___/_____ Aeration Begun ___/___/_____ Aeration Completed ___/___/_____

DURATION OF FUMIGATION: Total Hours Under Gas ________ Total Hours of Aeration ________

Time After Fumigant Placement (Hours)	6	12	24	48	72	96	120*	144*	End of Aeration
AIRSPACE Phosphine Conc. (ppm)									
CASE/BALE PROBE** Phosphine Conc. (ppm)									
AMBIENT TEMP (°C)									

* If required. ** The minimum requirement is 200ppm for 48 consecutive hours in a 96 hour period. The gas must be taken at the centre of the innermost case/bale.

COMMENTS

SIGNATURES

[Attesting the information entered above is correct, that the fumigation practices adhered to the consignees instructions and that all fumigant residues were removed from the consignment.]

SUPERVISING FUMIGATOR: Name________________ Signature______________ Date:________

FOR THE SUPPLIER: Name________________ Signature______________ Date:________

8.3 CHECKLIST FOR A WAREHOUSE FUMIGATION WITH MAGNESIUM PHOSPHIDE PLATES: A PRACTICAL EXAMPLE

Pre-fumigation

1. Read, understand and follow the fumigant label (and instruction manual) in accordance with local laws and ordinances. Verify fumigators have all necessary licences and training.
2. Plan for the fumigation in coordination with leaf storage personnel, fumigators (possibly a contractor), safety and quality assurance personnel. Ensure any contract holds the fumigator responsible for the seal and therefore any fumigation failures.
3. Notify police, fire fighters, or local medical facility of intent to fumigate.
4. Ensure security personnel are available with a list of persons authorized to approach the fumigation site and emergency response team members.
5. Verify that the sprinkler system is in good working order and that the power supply to the valve house will not be interrupted.
6. Prepare warning placards with emergency names and telephone numbers.
7. **SEAL the fumigation site, to prevent gas leakage.**
8. Calculate the volume of the fumigation site and therefore the amount of fumigant needed. One plate releases 33 g phosphine: 1 strip releases $20 \times 33 = 660$ g phosphine: recommended dose is 1 g/m^3: supposed volume of warehouse is 15,200 m^3: fumigant needed = $15,200 \div 33 = 460$ plates or $15,200 \div 660 = 23$ strips.
9. Provide material safety data sheet; to be taped to the fumigation site and in the possession of the fumigator, security and safety personnel.
10. Assemble equipment to include: respiratory protective gas masks and SCBAs; phosphine monitoring devices; suitable fire extinguishers (no water); water drums for wet deactivation; locks and keys to secure the fumigation site.
11. Position tubing for monitoring phosphine concentrations in the airspace and tobacco.
12. Measure ambient/tobacco temperatures and obtain weather forecasts. Ensure the fumigant is not placed or removed in fog or rain; also it is not advisable to fumigate if strong winds are expected.
13. Communicate potential hazards of phosphine to all involved personnel.
14. Mark the position(s) where fumigant will be placed.
15. Mark the restricted area perimeter of at least 20 m.
16. Make last insect count and remove all pheromone traps.

Fumigation

1. Clear all non-essential personnel from the fumigation site, inspect the site to be sure and lock all entrances but one, to be used as the fumigators' exit.
2. Place the fumigant to expose both surfaces, moving toward the selected exit.
3. Seal, lock and label the exit when all fumigators have left the fumigation site.
4. Monitor gas concentrations at 6, 12, 24, 36, 48, 60, 72, 96 and every further 24 hours until completion.
5. Post a guard for the entire fumigation duration and have the fumigator present for the first 24 hours.
6. Check the structure and surrounds for possible leaks and repair, wearing necessary respiratory protection.
7. Add additional fumigant, if necessitated due to a leak, by fumigators entering with SCBAs if local law allows. This should only be considered if absolutely necessary and must follow stringent confined-space entry procedures. This is a potential hazard and should not be necessary. Rather improve the gas-tightness of the structure by additional sealing.
8. Continue the fumigation for **at least 96 hours.**

Post-fumigation

1. Open doors and vents, using respiratory protection, from the outside or partially raise the sheet.
2. Allow to aerate until the airspace is less than 0.1 ppm phosphine. Readings may be made by workers wearing respiratory protection; 2 days is often adequate. Respect the local laws relating to environmental release rates of phosphine.
3. Continue aeration until the tobacco mass is less than 0.1 ppm phosphine; usually 4 days suffices. However, if polyethylene liners are used as tobacco wrappers, this may require up to 6 weeks.
4. Remove the fumigant residues (spent plates), preferably when the airspace is less than 0.1 ppm, otherwise using respiratory protection.
5. Move the spent fumigant in *open* receptacles to a placarded, secure, ventilated, on-site wet deactivation site.
6. Place a maximum of 240 plates into a 110-litre mesh basket and immerse in 200 litres of water in a drum. Ensure all plates are covered by water at all times. Leave to deactivate all the unspent magnesium phosphide; usually 2 days will suffice.
7. Dispose of deactivated plates in an authorized landfill or burn. Discharge the water into a sewer, if available, or on to the ground. Obey local environmental regulations.

8. Release the warehouses and tobacco when both phosphine levels are below 0.1 ppm.
9. Re-start pheromone trap monitoring for *L. serricorne* and *E. elutella* with new traps.
10. Review the fumigation and optimize procedures for the next time.

8.4 METHYL BROMIDE

Methyl bromide (CH_3Br) or bromomethane has the following physical properties: molecular weight, 94.9; boiling point, 3.6°C; solubility in water (at 20°C), 1.75 g/100 g; specific gravity, 1.732; melting point, –69.8°C. It is a colourless, odourless gas at normal temperatures and pressures and non-flammable in air. Although generally non-corrosive it will degrade natural rubber, react with aluminium to produce spontaneous ignition from aluminium trimethyl, and may produce off-taste from multiple fumigations.[44] Methyl bromide was first used in 1932, available in steel, pressurized cylinders initially with a warning agent (2% chloropicrin). Only 100% methyl bromide should be used because of the hazardous nature of chloropicrin. Usually, the speed of a vacuum fumigation is the factor that decides its use.

Because of this potential off-taste and the potential risks involved with this fumigant, its use is decreasing in the industry. When used, fumigate tobacco once and always use trained, certified, professional fumigators. Methyl bromide, as an ozone depleter, is regulated under the Montreal accord[10] and the US' Clean Air Act. As such, it is in a world-wide phase-out period. Therefore, if this fumigant is part of your current control programme alternatives such as sanitation and phosphine should be found.

Methyl bromide penetrates poorly (less than 15 cm) into the tobacco mass under atmospheric pressure[170] where it is usually dosed at 32 g/m^3 at 7–20°C ambient or 20 g/m^3 at greater than 21°C ambient.[40,356] Different accepted treatment schedules are shown in Table 8.2.

Table 8.2 Accepted treatment schedules for methyl bromide fumigations in vacuum fumigation chambers

Dose (g/m^3)	Exposure (hours)	Temperature (ambient) (°C)	Airwashes
64	4	above 20	6
80	6	10–20	8
80	8	5–10	8

In general, the requirements for a methyl bromide fumigation are those for phosphine. Indeed, phosphine will be used for atmospheric fumigations. Therefore, we will address vacuum fumigation with methyl bromide. Methyl bromide does obey Haber's rule where: concentration of gas × duration = a constant.[40]

Because methyl bromide is heavier than air, specially constructed vacuum-fumigation chambers may be used, which allow air circulation, and a complete exchange of the airspace (an airwash). Ten airwashes are required if tobacco is within polyethylene-lined packaging to ensure adequate aeration. Such liners should be open to allow adequate penetration and aeration. An airwash requires the complete exchange of chamber gas with fresh air following a 5-minute period under a 94 kPa vacuum. Often the schedule will be 96 g/m^3 to allow for the difficulty in penetrating tobacco liners.

Respiratory protection must be worn at all times, as this gas is odourless and toxic. Wear loose protective clothing to avoid burns if the liquid is held against the skin; therefore, avoid rubber boots and gloves. Eye protection is mandatory. Tobacco is stacked with greater than 3 cm between stacks to allow circulation.

Continue aeration of the airspace and tobacco until less than 0.1 ppm before moving the tobacco. Measure gas levels during and after the fumigation via detection tubes.

In the European Union the maximum tobacco residue levels for bromide residues are 50 ppm and 0.1 ppm for methyl bromide.

Briefly, the procedure for vacuum-chamber fumigation is as follows:

1. Load the chamber with tobacco; close and seal the door.
2. Reduce the pressure in the chamber minus 94 kPa.
3. Turn on the gas circulation fan(s); fan(s) should operate throughout the fumigation.
4. Check chamber seals and condition by monitoring the pressure.
5. Open valve and introduce the correct dose of liquid/gas.
6. After the correct exposure (4–8 hours), pump gas out of the chamber and thus return to normal atmospheric pressure, by allowing fresh air to enter.
7. Repeat step 2.
8. Pump gas out of the chamber and thus return to normal atmospheric pressure, by allowing fresh air to enter.
9. Repeat steps 2 (7) and 8 – and airwash the correct number of times (6–10).
10. Check the gas concentration in the chamber and repeat step 9 until less than 0.1 ppm is present in the tobacco and airspace.
11. Open chamber and remove tobacco.

8.5 CONTROLLED ATMOSPHERES

Low oxygen (O_2) or high carbon dioxide (CO_2)[84] concentration atmospheres (or a combination of both) will suffocate and/or dehydrate insects infesting stored tobacco. This began as hermetic storage in 1000 BC and is experiencing a revival. Atmospheres may be controlled by: adding nitrogen or carbon dioxide to the airspace; burning hydrocarbons in air and using the (so-called) controlled atmosphere (CA) generator's exhaust as a purge; removing the oxygen using an oxygen scavenger. A vacuum is not effective since even a high vacuum leaves the air in the same proportions of N_2:O_2 of 5:1 and insects require only small amounts of oxygen[168]. Terminology in this field is a little confused as controlled atmospheres could equally apply to all fumigants. Modified atmospheres is meant to differentiate a single event fumigation (one-shot) from the continuous controlled fumigation. "Inert" atmosphere fumigation is a misnomer as these atmospheres are lethal, not inert.

Controlled atmospheres of greater than 60% CO_2 or less than 1% O_2, or a combination are in use today. Some Italian cigarette manufacturers use the Isolcell® CA generator system. This is a fumigation chamber with adjustable temperature and humidity attached to the Isolcell generator. Usually a 14-day treatment is required from loading the chamber to removal of the aerated tobacco. This generator and others are specialized equipment and details should be sought from the manufacturers. Generators and low oxygen atmospheres do not lend themselves to warehouse fumigations, since the seal will not usually maintain less than 1% O_2.

Large-scale CO_2 fumigations have been reported,[181,192] but are unlikely to find favour due to environmental concerns of releasing a greenhouse-effect gas, and safety concerns surrounding the aeration of large volumes of CO_2. However, 60% CO_2 at 28°C and 65% RH for 8 days will kill 100% of infesting *L. serricorne*. Work is underway to evaluate small-scale fumigation chambers with and without a vacuum system. Other novel uses include high-pressure CO_2 treatments.[157,158] Good fumigation practices are necessary for these fumigations. SCBAs are mandatory as gas masks offer no protection to oxygen deficiency. Monitoring gas concentrations with electronic devices and/or reactive tubes is essential, as is knowing the temperature and humidity. Again, difficulties will be encountered when polyethylene liners are used.

The above fumigants are experimental and/or little-used in the tobacco industry. However, the food industry uses them extensively for grain fumigations (e.g. Australia since 1970) and modified atmosphere packaging. Further information is available, and for those considering their use, a competent experienced, professional fumigation company should be consulted.

Case histories 9

The following incidents demonstrate the infestation control problem-solving process. Each demonstrates a system failure of the IPM programme or problems encountered during IPM programme implementation. Only eight examples are given, as detailing all such possible system failures would require a second volume.

9.1 SPOTTING

Problem scenario

Inspection of the visual quality of cigarettes at the maker revealed a high level of spotting of the cigarette paper. PRODUCTION STOPPED.

Investigation

Severity of the spots was quantified: the percentage of cigarettes affected; the individual and accumulative size of spots.

Source assessed: the makers involved; the batch of cut filler/tobacco blend used.

Affected product pinpointed: visual quality analysis of cigarettes produced prior to incident.

Visually inspected for cause of spots.

Results showed spotting exceeded product standard/specification on all machines for the whole of 3 days' production. Spots were associated with dead (95%) and living *L. serricorne* larvae: caused by larval body oils. These spots darkened on cigarette storage and are permanent.

Investigation showed massive *L. serricorne* infestation in tobacco storages. PRODUCTION REMAINED STOPPED.

Successful solution strategy

Production was halted for 2 days for intensive cleaning of the production line, and re-started with an alternate non-infested tobacco supply.

3-days non-standard cigarette production and ripper shorts were destroyed. This could have been fumigated, ripped, sieved (see below), and blended back in as rippers. Marketing's decision was to destroy.

Affected warehouses were closed for 8 days for phosphine fumigation of the entire tobacco stock.

The affected tobacco was re-introduced after being fumigated and blended with new stocks. Sieves were introduced to remove dead insect bodies. Sieves of 6/cm gave acceptable reduction of spots and increase of waste. Placement of the sieves was successful at pre-ordering cylinder and/or post-cutters. IPM was rigorously implemented to avoid re-occurrence.

Note: if this had been a cut filler/make-pack operation with infested cut filler as the problem, the same solution would apply. However, if logistics allowed, an alternate solution was possible by freezing the cut filler.

9.2 RE-INFESTATION

Problem scenario

Four weeks after a successful, whole warehouse, phosphine fumigation, the warehouse supervisor noted that pheromone trap counts had increased from zero to above 200 *L. serricorne* per trap overnight. He stopped counting at 200 as the trap's glue surface was black with beetles (Plate 3.7). He closed the warehouse, stopped all tobacco movement and called for help.

Investigation

Determined if other warehouses were involved via trap inspections.

Identified suspect tobacco lots – those which entered the warehouse over the previous 3 days.

Inspected suspect lots visually for infestation (section 3.1).

Reviewed factory/company/country-wide infestation situation and pheromone reports.

Results

One lot of 2,400 bales of oriental tobacco entered the warehouse the day before. It was not inspected on arrival as shipping documents indicated. Visual inspection revealed it to be severely infested with *L.serricorne* and *E. elutella*. The infestation outbreak was limited to one warehouse.

Successful solution strategy

Warehouse was closed for 8 days for phosphine fumigation. New pheromone traps were installed post-fumigation. Leaf department and shipping department re-worked shipment documentation to avoid a re-occurrence.

Affected lot inspected and judged in-specification. As a precaution against spotting it was used in production only as 20% of that grade in the blend. **Note**: the preferred solution would be to return the entire consignment to the vendor as out-of-specification. Unfortunately, the shipment's history precluded this approach.

9.3 CONTAMINATION

Problem scenario

Incoming inspection of eight container loads of tobacco revealed a greyish powder on the surface of several bales in each container. Leaf department was alerted and the containers were re-loaded and parked pending investigation.

Investigation

Container/lot history was ascertained with reference to fumigation standards.

Physico-chemical analysis was done on the suspect powder, looking particularly for phosphide/phosphine.

Results

The powder was the residue from aluminium phosphide fumigant tablets. Unfortunately, while in storage at the vendor's warehouse, improper fumigations were conducted with tablets placed on the bale surface. Residues were mostly aluminium hydroxide plus aluminium phosphide. The latter released phosphine on contact with water. Traces were found on the tobacco beneath the residues.

Successful solution strategy

Entire lot/shipment was rejected as contaminated and returned to vendor at their (owners') expense. The vendor was reminded of correct fumigation standards.

9.4 PHYTOSANITARY CERTIFICATES

Problem scenario

Incoming inspection of 20 container loads of tobacco revealed four containers with infested flue-cured tobacco. The infestation was detected initially by pheromone traps placed in-transit in the containers and verified by visual inspection. Affected containers were set aside.

Investigation

Ensured that phytosanitary certificate had been issued for all 20 containers by relevant government personnel at country of origin.

Ensured that contractual fumigation standard had been met by vendor personnel or fumigation contractor.

Investigated container history for possible re-infestation post-fumigation, for example during inspection/weighing of the tobacco at port of entry.

Results

Fumigation standard was not met, no gas readings were taken during fumigation, but phytosanitary certificates were issued. Therefore, failure of fumigation was suspected as there was no evidence of in-transit re-infestation.

Successful solution strategy

All four containers were re-fumigated at the vendor's cost and a written reminder was sent of the contractual fumigation standard.

9.5 ACQUISITION

Problem scenario

A recent acquisition in a temperate-climate country included a cigarette manufacturing factory and a multi-storey tobacco warehouse at 70% capacity. Installation of pheromone traps revealed an unacceptable level of *E. elutella* infestation.

Investigation

Inspection of the warehouses for suitability for whole-warehouse or stack fumigations. Inspection of all tobacco for location of infestation using pheromone trap data as guidance. Identified suitable pyrethrum/pyrethroid formulations, application equipment, and competent pest control contractors, in-country, for possible control effort.

Results

Numerous tobacco lots showed low-level infestation. A few tobacco lots had dense silk webbing on outside and heavy infestation within – these lots were several years old and no longer included in blends (Plate 3.3). The warehouse is on the factory site, only 5 m from the primary processing area and 7 m from residential areas. Floors are wooden and the roof is of a non-gas-tight construction. Therefore, fumigation, the solution of choice was not feasible. Deltamethrin formulations were available and labelled for use and two experienced pest control companies were willing to bid for the work.

Successful solution strategy

During the following week-end a contractor treated the whole warehouse with deltamethrin as a residual application to the walls and ceiling followed by a cold-aerosol (fog) treatment. After aeration new pheromone traps were placed and indicated a need for an additional treatment after 8 weeks. Monitoring re-commenced after the winter and continues to provoke necessary control treatments.

All of the tobacco was removed from the hotter "attic" floor of the warehouse. The heavily infested lots were removed and transported to a distant farmer for use as an agricultural supplement. Warehouse windows, doors and openings were closed or screened to prevent the entry of feral *E. elutella*.

9.6 "HIDE AND SEEK"

Problem scenario

Six months after installing/beginning the Factory Sanitation Programme (FSP), the production supervisors felt that everything had been done according to standards, and yet the pre-blend silo area in factory no. 2, north side continued to show below standard *L. serricorne* pheromone trap catches. They wrestled with this for several weeks, to no avail, and called for help.

Investigation

Reviewed pheromone trap data in relation to FSP and additional efforts, including mobile trap data.

Visually inspected the affected area with appropriate area personnel, during a machinery down-time or out-of-shift to enable in-depth inspection.

Ensured that area/machine supervisors realized we were seeking beetles, but they would be "hiding" in food. Since their food is tobacco, we looked first for anywhere tobacco and/or tobacco dust may have accumulated inside or outside processing equipment.

Results and solutions

No obvious source of infestation or harbourage was found. Factory personnel had maintained the FSP standard. Mobile traps had delineated the affected area but this was a rather simple cube containing equipment. The search for tobacco dust was more revealing: three locations with tobacco dust accumulations with *L. serricorne* were found.

These were:

- an extractor duct branch had been closed. Unfortunately, instead of a flush finish, a truncated Y had been left and 20 kg of dust had accumulated in the truncated branch – this was altered to be flush.
- an "hermetically" sealed air-separator. Here, on removing a side-panel a void was found – a useless area that formed a 2 × 1.5 × 1-m pocket. This pocket was full of dust and *L. serricorne*. The remedy involved the structural alteration of the machine; a short-term solution was a door for cleaning access.
- a 30-cm pipe had been removed. Where it traversed a wall, the hole was "sealed" by two metal plates screwed in place. Removal revealed not only dust and beetles in the hole but also in the exposed wall cavity. Pyrethroid (permethrin) was dusted into the wall void after cleaning as much as possible, and the hole closed with bricks and mortar. An insect growth regulator would have been preferred to permethrin; however, the former was not registered for use in-country.

9.7 CONSUMER COMPLAINTS

Problem scenario

The consumer relations department head sent four partial packs of cigarettes, that were punctured and damaged. For a detailed report his questions were: 1. Are insects involved; 2. If so, which ones; 3. If so, did they originate in a production facility?

Investigation

Examined all cigarettes with a 20-fold hand lens and 100-fold binocular reflectance microscope. Their appearance was drawn and observations noted, including production code details, etc.

Lightly tapped the cigarettes, with the hole downwards, over white paper and examined particles that were tapped out.

Slit the cigarettes open and examined, microscopically, the tobacco and inner surface of the cigarette paper.

Noted all findings in a controlled notebook in case of litigation, and took photographs of critical findings.

Translated the pack or cigarette codes into real time, date, machine, factory, etc.

Results of examination made on 22.06.1994:

Cigarette no. 1: code CC00347, manufacture date 14.03.1994, shift 2, machine L94, factory XYZ.

Hole: typical of a piece of stem piercing the paper outwards.

Successful solution strategy

Reviewed visual quality audit data for factory XYZ. If stem holes are trending towards out-of-specification, initiate corrective action on the stem process.

Cigarette no. 2: code CC00348, manufacture date 03.05.1994, shift 1, machine P72, factory ABC.

Hole: typical appearance of insect bite marks (Plate 9.1) with excrement smears on the tipping paper, most likely cockroach damage.

Successful solution strategy

Advised the consumer, via consumer relations department, that the damage was caused by an insect, probably a cockroach, that entered the pack once it was opened.

Cigarette no. 3: code CC00349, manufacture date 06.04.1994, shift 3 machine P03, factory NMO.

Hole: typical appearance of paper fault with the hole edges transparent. The paper thickness increases to normal, progressively, away from the hole.

Successful solution strategy

Shared these findings with the supplier, reviewed the suppliers' quality rating, especially in relation to paper holes. Took quality improvement action based on these findings.

Cigarette no. 4: code CC00350, manufacture date 13.10.1993, shift 2 machine HL94, factory XYZ, returned from semi-tropical climatic zone.

Hole: typical of *L. serricorne* damage (Plate 9.2), small faecal pellets (frass) were tapped out of the hole during examination (Plate 9.3). Cigarette paper showed that the beetle emerged from within the cigarette as the inner surface was rough from chewing (Plate 9.4), whereas the outer surface was smooth (Plate 9.5).

Plate 9.1 Holes in the cigarette paper caused by a cockroach (*Peripleneta americana*) bite. Simulated to mimic damage from consumer-collected cigarettes.

Plate 9.2 Holes in cigarette caused by *L. serricorne*.

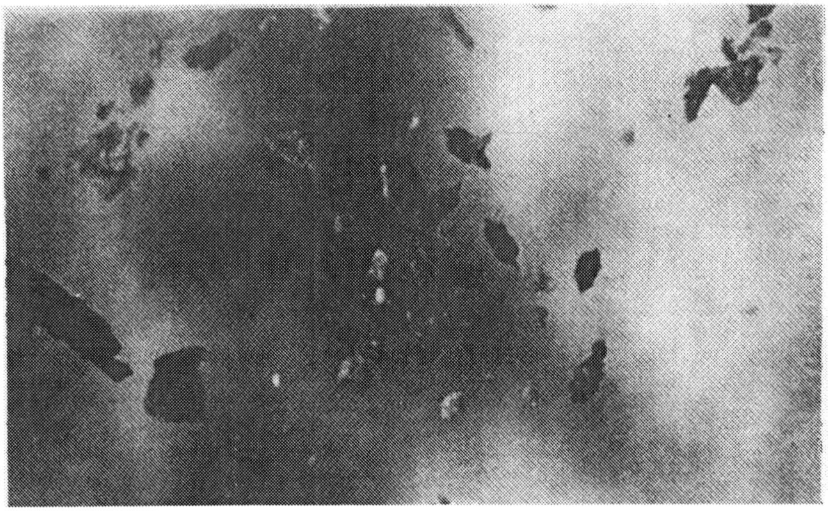

Plate 9.3 Frass or faeces of *L. serricorne*; see also Plate 2.6.

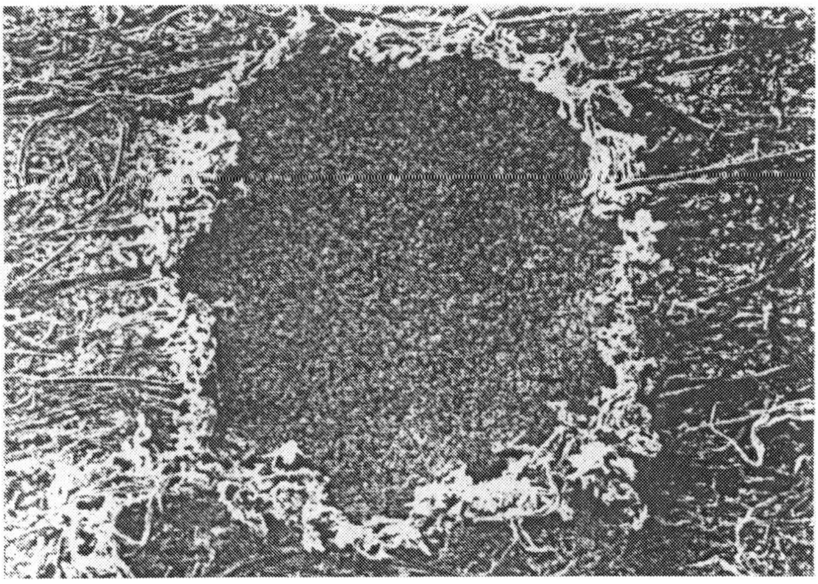

Plate 9.4 *L. serricorne*-caused hole in a cigarette paper (Plate 9.2): the inner surface of the cigarette paper (see Plate 9.5 for the outer surface).

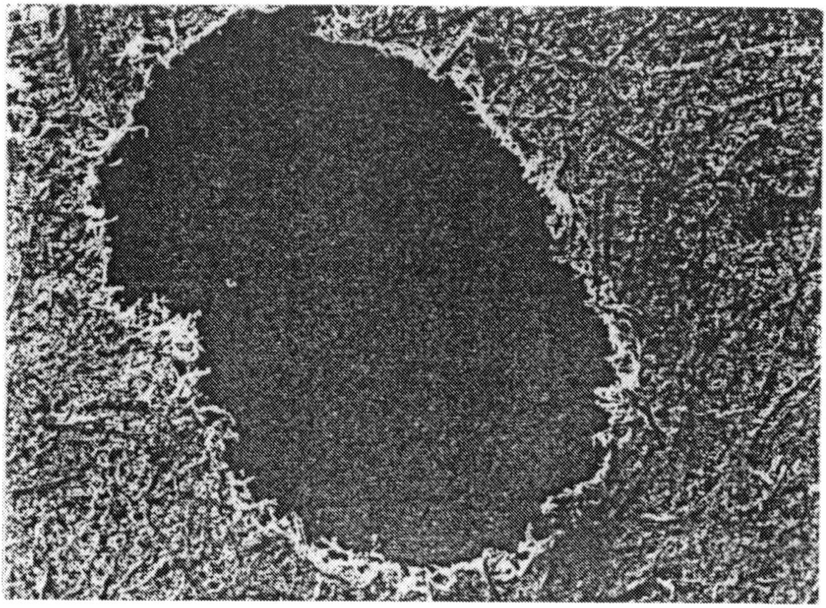

Plate 9.5 *L. serricorne*-caused hole in a cigarette paper (Plate 9.2): the outer surface of the cigarette paper (see Plate 9.4 for the inner surface).

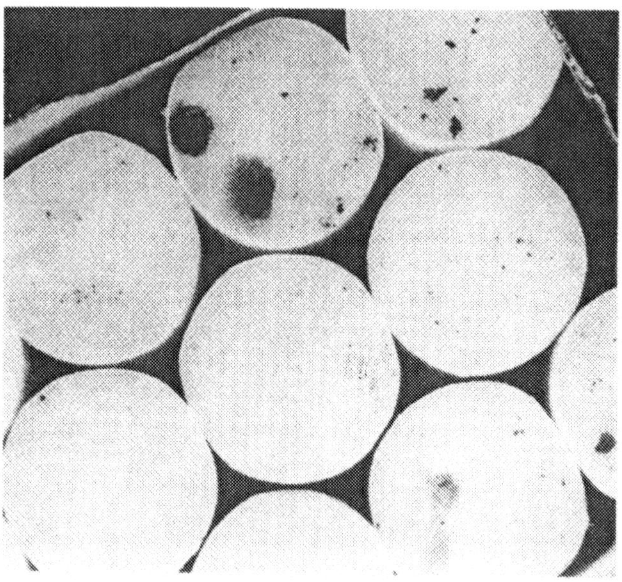

Plate 9.6 Cigarette filter showing burrow holes made by *L. serricorne*.

Eight larvae were found in other cigarettes in the pack, one burrowing through the tow (Plate 9.6).

Two empty pupal cases were found, one surrounding one of the holes in the paper.

Three dead adults were found within the pack.

One exit hole was found through the foil, pack, and outerwrap (Plate 9.7). These were known to be exit holes from the same considerations, as mentioned above, of roughness on the chewed inner surface.

Review of *L. serricorne* monitoring data from the factory XYZ showed no insects were detected in cut rag storage areas or in making areas for the previous 27 months.

Conclusion

The cigarettes were 8 months old. *L. serricorne* generation time is 7 to 8 weeks (section 2.1.2) and only one (or two) generations had been completed in the pack. Therefore, the cigarettes were infested in the retail channel, most likely at the point of sale. *L. serricorne* entered the pack via the pack seal, by crawling through the join as a first instar larvae, hence, no entry hole is ever likely to be found.

Successful solution strategy

Implemented corrective action to ensure cigarettes are in the market for no more than the marketing specification, which was breached in this case.

Plate 9.7 Burrow hole of *L. serricorne* exiting the cigarette pack via the foil, pack and outerwrap.

Emphasized sanitation to all retail outlets, especially the one concerned.

Apologized and explained the situation to the consumer. Entomophobia should not be underestimated.[176]

9.8 TO CLEAN OR NOT TO CLEAN.....?

Problem scenario

The infestation level at one of "our" factories was increasing and already above acceptable levels. The factory management team were devoting resources to emergency ("fire-fighting") cleaning and the usual pesticide usage. They felt that recommendations given had not worked, they asked for help, and talked seriously about fumigating the factory!

Investigation

In collaboration with the factory management team, the annual cost for sanitation activities and pesticide application contract costs was calculated. Sanitation costs included contractors' and factory personnel's time, the latter calculated as the gross cost.

Using the existing Factory Sanitation Programme (FSP, section 4.2.4) factory production areas were inspected and audited. Places that had been missed were identified and subjected to emergency cleaning.

Successful solution strategy

Updated the FSP and re-defined the cleaning schedule based on the current budget established during the investigation. Re-distributed the time from redundant or excessive cleaning to the areas that were missed.

Ensured that adequate vacuums and attachments were available.

Initiated training for all personnel: stressing the idea of "**clean it right the first time**".

Ensured contractors were supervized by production personnel.

Phased out insecticides as the FSP obviates them.

Explained the impossibility of a factory fumigation (Chapter 8).

Appendix A: Equipment suppliers

Mention of a company or product does not constitute a recommendation, nor is this list exhaustive. These offices may be contacted for local distributors.

PHEROMONE TRAPS FOR **L. serricorne**

NEW SERRICO TRAPS:
FUJI FLAVOR Co. Ltd.,
3-5-8 Midorigaoka, Hamura,
TOKYO, 190-11, JAPAN.
Tel: (425) 55 5100
Fax: (425) 55 7920

INSECTS Ltd.,
10540, Jessup Blvd.,
INDIANAPOLIS,
IN 46280-1451, USA.
Tel: (1) 317 846 3399
Fax: (1) 317 846 9799

MARKETBELLS Co. Ltd.,
828 Ocean Centre,
Harbour City,
TSIMSHATISUI, KOWLOON, HONG KONG.
Tel: (852) 7361 733
Fax: (852) 7302 856

MACHINERY & CHEMICALS
St Ivel Way,
Warmley,
BRISTOL BT15 5TY, UK.
Tel: (117 9) 600 060
Fax: (117 9) 352 437

LASIOTRAP
H. BORGWALDT GmbH.
Schnackenburgallee 15,
22525 HAMBURG, GERMANY.
Tel: (49) 40 85 31 38-0
Fax: (49) 40 850 56 00

PHEROMONE TRAPS FOR **E. elutella**

STORGARD
INDUSTRIAL PESTICIDES
7-29 Brasenose Road,
LIVERPOOL L20 8H4, UK.
Tel: (44) 151 933 7292
Fax: (44) 151 922 3733

TRÉCÉ Inc.,
P.O. Box 6278,1143 Madison Lane
SALINAS, CA 93912, USA.
Tel: (1) 408 758 0204
Fax: (1) 408 758 2526

MOTH TRAP
DELTA TRAP
AGRISENSE - BCS Ltd.,
Treforest Industrial Estate,
Pontypridd,
MID GLAMORGAN CF 375 SU, UK
Tel: (44) 1443 841 155
Fax: (44) 1443 841 152

MAS TRAP
AGRIMONT
Montedison Group,
Piazza della Republica, 14/16,
20124 MILANO, ITALY.
Tel: (39) 262 701 - 6333.1
Fax: (39) 26333.2018

MULTI-PHER
2949 Ch. Ste-Foy,
STE-FOY, QUEBEC G1X 1P3,
CANADA.
Tel: (1) 418 653 3101

SANITATION EQUIPMENT

TORCHES (FLASHLIGHTS)
MAAG INSTRUMENTS,
1653, S. Sacramento Ave.,
ONTARIO,
CA, 91761-7769, USA.
Tel: (1) 714 947 1006
Fax: (1) 714 947 3116

INDUSTRIAL VACUUM CLEANERS AND ATTACHMENTS
AKKERMANS for NARVIK®,
Nijverheidsweg, 17,
4731, OUDENBOSCH, HOLLAND.
Tel: (31) 1652 13 555
Fax: (31) 1652 17110

KARCHER,
1030 Villars St. Le Croix,
LAUSANNE, SWITZERLAND.
Tel: (41) 21 35 55 46

NUMATIC Int. Ltd.,
BEAMINSTER, DORSET, DT83PR, UK.
Tel: (44) 1308 862 062
Fax: (44) 1308 862 766

RUWAC SCHIESTL GmbH,
6424 SILZ, 9, AUSTRIA.
Tel: (43) 5263 6347
Fax: (43) 5236 6638

WAP Int. Inc.,
170, E. Freedom Ave.,
ANAHEIM, CA, 92801, USA.
Tel: (1) 714 871 8500
Fax: (1) 714 871 6895

PHOSPHINE FUMIGANTS

DETIA DEGESCH GmbH,
Dr.-Werner-Freyberg Str. 11,
69514 LAUDENBACH / BERGSTR.,
GERMANY.
Tel: (49) 6201 7080
Fax: (49) 6201 708 402

SUNZON INT. Ltd.,
5511 Capital Center Drive,
Suite 302,
RALEIGH, N.C. 27602-3365, USA.
Tel: (1) 919-859 2500
Fax: (1) 919-859-2155

PHOSPHINE DETECTION AND SAFETY EQUIPMENT

AUER MSA (Suisse) SA,
Grubenstr. 24,
3072 OSTERMUNDIGEN 1,
SWITZERLAND.
Tel: (41) 31 932 1151
Fax: (41) 31 932 1132

DETECTAWL (GASTEC),
Unit 61, Garamonde Drive,
WYMBUSH, MILTON KEYNES,
BUCKS MK8 8DE, UK.
Tel: (44) 1908 568 076
Fax: (44) 1908 260 593

DRAGERWERK,
Postfach 1339, Moislinger Allee 53-55,
23558 LUEBECK 1, GERMANY.
Tel: (49) 451 8820
Fax: (49) 451 882 2080

DRAEGER-BRANDT SA,
3c Route de la Fédération,
BP No. 141,
67025 STRASBOURG CEDEX,
FRANCE.
Tel: (33) 88 39 18 47
Fax: (33) 88 79 28 21

KOMYO RIKAGAKU KOGYO K.K.,
(KITAGAWA),
8-24 Chuo-cho,
1 Chome,
Meguro-Ku,
TOKYO 152, JAPAN.
Tel: (81) 3 570 43 511
Fax: (81) 3 570 43 316

BEDFONT INSTRUMENTS FOR
ELECTRONIC DETECTORS,
Bedfont House, Holywell Lane,
Upchurch,
SITTINGOURNE, KENT ME9 7HN,
UK.
Tel: (44) 1634 375 614
Fax: (44) 1634 378 980

G.D.,
ZA DU VERT GALANT,
12, Rue des Oziers,
95310 SAINT-OUEN L'AUMONE,
FRANCE.
Tel: (33) 1 34 64 68 88
Fax: (33) 1 34 64 32 87

MST GmbH
P.O. Box 16, Benediktstr. 1,
82069
HOHENSCHAFTLARN / MUNICH,
GERMANY.
Tel: (49) 8178 8690
Fax: (49) 8178 8220

WEATHERALL EQUIP. &
INSTRUMENTS Ltd.,
P.O. Box 69,
Tring,
HERTFORDSHIRE HP23 6PL, UK.
Tel: (44) 1494 758 110
Fax: (44) 1494 758 014

PHOSPHINE DETECTION AND SAFETY EQUIPMENT (continued)

CABOT SAFETY Ltd.,
First Avenue,
Poynton,
STOCKPORT, CHESHIRE SK12 1FJ, UK.
Tel: (44) 1625 87 83 20
Fax: (44) 1625 87 73 48

SABRE SAFETY Ltd.,
Matterson House,
HAMPSHIRE GU12, UK.
Tel: (44) 1252 34 41 41
Fax: (44) 1252 219

RACAL SAFETY Ltd.,
WARRINGTON, CHESHIRE WA5 5NY, UK.
Tel: (44) 1925 234 656
Fax: (44) 1925 234 662

MARTINDALE PROTECTION Ltd.,
Neasden Lane,
LONDON NW10 1RN, UK.
Tel: (44) 181 450 8561
Fax: (44) 181 450 4794

PYRETHRIN / PYRETHROID FORMULATIONS

FROWEIN GmbH,
Am Reislebach 83,
72461 ALBSTADT (TAILFINGEN),
GERMANY
Tel: (49) 7432 3044
Fax: (49) 7432 8944

AGREVO
ENVIRONMENTAL HEALTH,
McIntyre House, High Street,
BERKHAMSTEAD, HERTS, HP4 2DY, UK.
Tel: (44) 1442 86 33 33
Fax: (44) 1442 87 07 61

METHOPRENE FORMULATIONS

SANDOZ AGRO SPECIALTY PRODUCTS Inc.,
1300, E. Touhy Ave.,
DES PLAINES, IL, 60018, USA.
Tel: (1) 708 390 3074
Fax: (1) 708 390 3940

SANDOZ SPECIALTY PEST CONTROL Ltd.,
SGS House, 217-221 London Road,
CAMBERLEY, GU15 3EY, UK.
Tel: (44) 1276 25 425
Fax: (44) 1276 25 769

INSECTICIDE APPLICATION EQUIPMENT

B & G EQUIPMENT Co.,
P.O. Box 130, Plumsteadville,
PA 18949, USA.
Tel: (1) 215 766 8811

COOPER, PEGLER & Co. Ltd.,
(Surface Sprayers)
283 London Road,
Burgess Hill,
WEST SUSSEY, RH15 9QU, UK.
Tel: (44) 1444 24 25 26
Fax: (44) 1444 23 55 78

CURTIS DYNA-FOG Ltd.,
17335 US Highway, 31,
N. WESTFIELD, IN 46074, USA.
Tel: (1) 317 896 2561
Fax: (1) 317 896 3788

KILLGERM,
P.O. Box 2,
Denholm Drive,
OSSETT, W YORKSHIRE WF5 9NB,
UK
Tel: (44) 1924 277 631
Fax: (44) 1924 264 757

LOWNDES ENGINEERING
(Leco Space Sprayers),
125 Beauchard Street,
P.O. Box 488, Valdosta,
GEORGIA, 31603-0488, USA.
Tel: (1) 912 242 3361

MICRO-GEN EQUIPMENT Corp.,
(Space sprayers),
10700 Sentinel Drive,
SAN ANTONIO, Texas 78 217, USA.
Tel: (1) 654 8570

MONTAN GmbH, (Fontan Space
Sprayers),
P.O. Box 1260,
Max-Euth-Weg 42,
7972 ISNY IM ALLGAN, GERMANY.

ROUSSEL BIO Corp.,
400 Sylvan Avenue,
ENGLEWOOD CLIFFS, NJ 07423,
USA.
Tel: (1) 201 871 9845
Fax: (1) 201 871 0771

KABAT APPLICATION EQUIPMENT

W.H. DICKINSON ENGINEERING
Ltd.,
Moorside Road,
WINCHESTER, HANTS SO23 7SS,
UK.
Tel: (44) 1962 84 2222
Fax: (44) 1962 84 0567

CHEMTRADE INTERNATIONAL
Ltd.,
P.O. Box 237, South Esplanade,
ST. PETER PORT, GUERNSEY, UK.
Tel: (27) 441 770 118 (S. Africa office)
Fax: (44) 1481 72 8892

GBE LEGG,
Newbury Road,
ANDOVER, HANTS SP10 4DW, UK.
Tel: (44) 1264 58 611
Fax: (44) 1264 333 873

METHOPRENE DETECTION

MILLIPORE Corp.,
80 Ashby Road,
BEDFORD, MA 01730, USA.
Tel: (1) 617 275 9200
Fax: (1) 617 275 5550

ENTOMOLOGICAL EQUIPMENT

WATKINS AND DONCASTER,
P.O. Box 5, Cranbrook,
KENT TN18 SE2, UK.
Tel: (44) 1580 75 31 33
Fax: (44) 1580 75 40 54

PEST CONTROL CONTRACTORS

BRUCE TERMINIX Co.,
P.O. Box 14009,
GREENSBORO, N.C. 27405, USA.
Tel: (1) 919 274-3103
Fax: (1) 919 621 6216

PESTCON SYSTEMS, Inc.,
P.O. Box 399,
WILSON, N.C. 27894-0399, USA.
Tel: (1) 919 237 7923
Fax: (1) 919 243 1832

PESTICIDE SERVICES Co. Ltd.,
Room 1408, Block A,
Sea View Estate,
NORTH POINT, HONG KONG.
Serves Hong Kong and mainland China

RENTOKIL Ltd.,
Felcourt, East Grinstead,
WEST SUSSEX, RH19 2JY, UK.
Tel: (44) 1342 833 022
Fax: (44) 1342 32 62 29

PHOSPHINE FUMIGANTS

DETIA DEGESCH GmbH,
Dr.-Werner-Freyberg Str. 11,
69514 LAUDENBACH / BERGSTR.,
GERMANY.
Tel: (49) 6201 7080
Fax: (49) 6201 708 402

SUNZON INT. Ltd.,
5511 Capital Center Drive,
Suite 302,
RALEIGH, N.C. 27602-3365, USA.
Tel: (1) 919-859 2500
Fax: (1) 919-859-2155

Appendix B: Glossary

Anobiidae: a family of beetles, generally infesting dead or decaying wood, to which belongs *L. serricorne* and *Stegobium paniceum*.

Antenna: a pair of segmented, sensory appendages on either side of the head, commonly referred to as "feelers".

Aposymbiotic: without symbionts.

Cephalic glands: glands in the head.

Claspers: male genital equipment for holding the female during copulation.

Cuticle: insect "skin".

Diapause: a delay in development not the result of adverse conditions.

Electroantennogram: record of antennal sensor (nerve) response to a stimulus.

Elytra: the hardened protective forewings of beetles that cover the hindwings: the latter used for flight.

Instar: the stage between moults or growth stage.

Larva: sexually immature insect form that hatches from the egg and transforms to the pupa commonly referred to as the "grub", "worm", "caterpillar", or "maggot".

Mandibles: front pair of mouthparts or jaws.

Moult: shed an old cuticle to reveal a new "larger" cuticle that allows growth.

Oviduct: egg duct from the ovaries to vagina.

Oviposition: egg "laying" or deposition.

Palpi: sensory structures on some mouthparts.

Pheromone: external secretions that influence the behaviour and/or physiology of the same species.

Prothorax: first thoracic segment.

Pupa: inactive stage of transformation from larva to adult.

Pupal cell: chamber for pupal development.

Sclerotized: hardened cuticle, therefore, darker in colour.

Serrate: saw-like.

Symbiont: a mutually beneficial interdependent relationship between species.

Tarsal formula: the number of segments (tarsomeres) in the "feet" of the front, mid, and hind legs. The leg is usually (from the body) composed of the following segments: coxa, trochanter, femur, tibia, *tarsus*, and pretarsus.

Tobacco types:
Burley – an air-cured tobacco of low sugar and high alkaloid content.
Flue-cured or Bright or Virginia tobacco heated in curing barns at up to 75°C for 6 days, has high sugar and low alkaloid content.
Oriental – is sun-cured with low nicotine and average alkaloid content.
Also fire-cured or smoked tobacco and other aircured and sun-cured variations.

References

1 Ahmad, M. (1985) Report of Trials in Pakistan using Kabat Tobacco Protector. *Tob Int* 16–24, November 1, 1985.

2 Ahmed, K.N. and Khatun, M. (1988) *Lasioderma serricorne* (F.) A possible alternate host of *Anisopteromalus calandrae* (Howard) (Hymenoptera: Pteromalidae) in Bangaladesh. *Bangaladesh J Zool* **16**: 165–166.

3 Aitken, A.D. (1963) A key to the larvae of some species of Phyticitinae (Lepidoptera: pyralidae) associated with stored products, and of some related species. *Bull Ent Res* **54**: 175–188.

4 Akehurst, B.C. (1968) *Tobacco*. Longman, London, 551pp.

5 Al-Badry, E.A., Rizk, G.N. and Hafez, S.M. (1980) Biological studies of the predatory mite *Acaropsis docta* attacking stored product insects. *Mesopotamia J Agric* **15** 179–202.

6 Aldridge, W.N. (1990) An assessment of the toxicological properties of pyrethroids and their neurotoxicity. *Crit Rev Toxicol* **21**: 89–104.

7 Alfieri, A. (1931) Les insectes de la tombe de Touthankhamon. *Bull Soc R ent Egypte* **15**: 188–189.

8 Ali, A.D., Donia, A.R. and El-Sawaf, S.K. (1972) The influence of natural food on the development and reproductive rate of *Lasioderma serricorne* Fab. (Coleopt., Anobiidae) *Z ang Ent* **72**: 212–220.

9 Ambadkar, P.M. and Khan, D.H. (1989) Observations on the influence of crude leaf extracts of some plants on adult and larval forms of the cigarette beetle, *Lasioderma serricorne* F. (Coleoptera: Anobiidae). *J Anim Morphol Physiol* **36**: 99–114.

10 Anonymous (1992) Montreal Protocol Assessment Supplement. Methyl Bromide interim scientific, technology and economic assessment. *UNEP*, June 1992.

11 Anonymous, (1991a) UK Health and Safety Executive, Fumigation using Phosphine. Guidance Note CSIO, May 1991.

12 Anonymous, (1991b) UK Health and Safety Executive, Fumigations using Methyl Bromide (Bromoethane). Guidance Note CS12, May 1991.

13 Arbogast, R.T., LeCato, G.L. and van Byrd, R. (1980) External morphology of some eggs of stored-product moths (Lepidoptera: Pyralidae, Gelechiidae, Tineidae). *Int J Insect Morphol Embryol* **9**: 165–177.

14 Arrestier, R., Delon, R., Tancogne, J. (1985) Efficiency of methoprene (Kabat) in the control of *Lasioderma* in Tobacco. *Ann Tabac Res Ing* **19**: 115–122 (Efficacité du methoprene (Kabat) dans la lutte contre le Lasioderme du tabac).

15 Asean and Acair (1989) Suggested Recommendations for the Fumigation of Grain in the ASEAN Region. Part 1 Principles and General Practice. *ASEAN Food Handling Bureau, Kuala Lumpur, 13pp,* (1989) (ISBN 967–9932–11–7).

16 Ashworth, J.R. (1993a) The biology of *Ephestia elutella. J stored Prod Res* **29**: 199–205.

17 Ashworth, J.R. (1993b) The biology of *Lasioderma serricorne. J stored Prod Res* **29**: 291–303.

18 Back, E.A. (1939) The cigarette beetle as a pest of cottonseed meal. *J Econ Ent* **32**: 739–749.

19 Back, E.A. and Cotton, R.T. (1926). The Cadelle. *Dep Bull US Dept Agric* ***1428*** 41pp.

20 Banks, H.J. and Ripp, B.E. (1983) Sealing of Grain Storages for Use with Fumigants and Controlled Atmospheres, *3rd International Working Conference on Stored-product Entomology.* Manhattan, KS, USA, October 23–28, (1983), pp. 375–390.

21 Banks, N. (1917) New mites, mostly economic (Arach. Acas.) *Ent News* **28**: 193–199.

22 Barak, A.V., Burkholder, W.E. and Faustini, D.L. (1991). Factors affecting the design of traps for stored-product insects. *J Kansas Ent Soc* **63**: 466–485.

23 Bare, C.O. (1942) Some enemies of stored-tobacco insects, with biological notes. *J Econ Ent* **35**: 185–189.

24 Bare, C.O. (1948) The effect of prolonged exposure to high vacuum on stored-tobacco insects. *J Econ Entomol* **31**: 109–110.

25 Bare, C.O., Tenhet, J.N. and Brubaker, R.W. (1947). Improved techniques for mass rearing of the cigarette beetle and tobacco moth. *USDA Bur Ent ET* **247**: 11pp.

26 Bare, C.O., Tenhet, J.N., Reed,W.D. (1946) Effect of the thermal-vacuum process on insects in stored tobacco. *J Econ Entomol* **39**: 612–613.

27 Beard, J.T., Childs, D.P., Fletcher, L.W., Joost, T.E., Arthur, J.H., Fletcher, D.G. (1986) Cooling tobacco in warehouses to kill cigarette beetles, Part III: passive solar and ventilation cooling system development. *Tob Sci* **30**: 1–10.

28 Beard, J.T., Iachetta, F.A., Childs, D.P., Fletcher, L.W. (1983) Cooling tobacco in warehouses during the winter to kill cigarette beetles, Part II. Model analysis of solar-driven refrigeration. *Tob Sci* **27**: 135–141.

29 Beesley, J.E. and Chadwick, P.R. (1986) The Activity of Insecticides on Surfaces. 7th British Pest Control Association Conference (Guernsey) 1986. 15 pages.

30 Beirne, B.P. (1952) *British Pyralid and plume moths.* Frederick Warne, London and New York, 208 pp.

31 Bell, C.H. (1975) Effects of temperature and humidity on development of four pyralid moth pests of stored products. *J stored Prod Res* **11**: 167–175.

32 Bell, C.H. (1976) Effect of cultural factors on the development of four stored-product moths. *J stored Prod Res* **12**: 185–193.

33 Bell, C.H. (1976a) Factors governing the induction of diapause in *Ephestia elutella* and *Plodia interpunctella. Physiol Ent* **1**: 83–91.

34 Bell, C.H. (1976b) Factors influencing the duration and termination of diapause in the warehouse moth, *Ephestia elutella*. *Physiol Ent* **1**: 169–172.

35 Bell, C.H. (1979) The Efficiency of Phosphine Against Diapausing Larvae of *Ephestia elutella* (Lepidoptera) Over a Wide Range of Concentrations and Exposure Times. *J stored Prod Res* **15**: 53–58.

36 Bell, C.H. (1981) The influence of light cycle and circadian rhythm on oviposition in five pyralid moth pests of stored products. *Phys Ent* **6**: 231–239.

38 Bell, C.H. (1983a) The effect of high temperatures on larvae of *Ephestia elutella* (Lepidoptera: Pyralidae) in diapause. *J stored Prod Res* **19**: 153–157.

39 Bell, C.H. (1983b) The regulation of development during diapause in *Ephestia elutella* (Hübner) by temperature and photoperiod. *J Insect Physiol* **29**: 485–490.

40 Bell, C.H. (1988) Minimum concentration levels of methyl bromide required for full efficacy against seven species of stored-product beetle at two temperatures. *Pest Sci* **24**: 97–109.

41 Bell, C.H. (1989) Influence of natural and laboratory conditions on the requirements for diapause in different breeding lines of the warehouse moth *Ephestia elutella* (Hübner). *J Insect Physiol* **35**: 137–141.

42 Bell, C.H. (1991) Diapause and cold tolerance of larvae of *Ephestia elutella*. *Postharv Biol Tech* **1**: 81–93.

43 Bell, C.H. (1992) Time, concentration and temperature relationships for phosphine activity in tests on diapausing larvae of *Ephestia elutella* (Hübner) (Lepidoptera: pyralidae). *Pestic Sci 35*: 255–264.

44 Benezet, H.J. (1989) Chemical control of pests in stored tobacco. *Proc 43rd TCRC*, Oct 2–5, 1989, Richmond, Va, pp.25.

45 Benezet, H.J. and Helms, C.W. (1986) A reduced rate of methoprene to control the cigarette beetle and its damage on stored tobacco. *Tob Sci* **30**: 130–131.

46 Benezet, H.J. and Helms, C.W. (1985) Reproductive potential of cigarette beetle populations exposed to methoprene. *Tob Sci* **29**: 14–15.

47 Benezet, H.J., Bowen, M.R., Helms, C.W. and Huffman, B.B. (1986) Toxicity of selected insecticides to *Lasioderma serricorne* (F) (Coleoptera: Anobiidae) by three methods of exposure. *J stored Prod Res* **22**: 93–96 .

48 Benezet, H.J., Helms, C.W., Rodgers, M.R. and Huffman, B.B. (1989) Toxicity of a new synthetic pyrethroid tralomethrin, to the cigarette beetle. *Tob. Sci.* 33: 34–36.

49 Benezet, H.J., Huffman, B.B. and Helms, C.W. (1988) Comparative toxicity of selected insecticides to the cigarette beetle at different temperatures. *Tob Sci* **32**: 41–43.

50 Benson, J.F. (1973) The biology of Lepidoptera infesting stored products, with special reference to population dynamics. *Biol Rev* **48**: 1–26.

51 Bodkin, G.E. (1914) Entomological Notes. *J Bd Agric Brit Guiana* **7**: 200–202.

52 Bond, E.J. and Miller, D.M.A. (1988) New Technique for Measuring the Combustibility of Gases at Reduced Pressures and Its Application to the Fumigant Phosphine. *J stored Prod Res* **24**: 225–228.

53 Bond, E.J., Dumas, T. and Hobbs, S. (1984) Corrosion of Metals by the Fumigant Phosphine. *J stored Prod Res* **20**: 57–63, 1984.

54 Boonkong, S., Leprayoon, C. and Meckvichai, W. (1986) Insects and mites found on stored garlic in Thailand. *Nat Hist Bull Siam Soc* **34**: 105–113.

55 Bousquet, Y. (1990) Beetles associated with stored products in Canada: An identification guide. *Agric Canada Publ* No. 1837, 220 pp.

56 Bovingdon, H.H.S. (1931) Pests in cured tobacco. The tobacco beetle, *Lasioderma serricorne* fab., and the Cacao moth, *Ephestia elutella* Hb., and aid to the recognition of the insects and the damage due to their activities. *Tobacco* **608**: 56–59.

57 Bridwell, J.C. (1920) Insects injurious to the Algaroba feed industry. *Haw Plant Rec* **22**: 337–343.

58 Brower, J.H. (1977) Interspecific matings between stored-product pyralidae (Phycitinae) *J Georgia Entomol Soc* **12**: 211–215.

59 Brower, J.H. (1979) Effects of irradiation on reproductive biology of adult tobacco moth, *Ephestia elutella. J Ga Entomol Soc* **14**: 201–209.

60 Bruce, W.A. and LeCato, G.L. (1979) *Pyemotes tritici* Potential biological control agent of stored product insects. *Recent advances in acarology*. Vol. 1 Ed. Rodriguez, J.G., Academic Press, New York.

61 Brunner, H.L. (1989) Grain preservation by means of refrigeration in tropical countries. *Sulzer Tech Rev* **4**: 19–23.

62 Buchelos, C.T. (1981) Coleoptera populations at flour mills and related areas. *Annls Inst phytopath Benaki* **13**: 6–29.

63 Buchelos, C.T. and Levinson, A.R. (1985) Population dynamics of *Ephestia elutella* (Huebner) in tobacco stores with and without insecticidal treatments: a survey of pheromone and unbaited traps. *Z ang Ent* **100**: 68–78.

64 Burgess, H.D. and Hurst, J.A. (1977) Ecology of *Bacillus thuringiensis* in Storage Moths. *J Inv Path* **30**: 131–139.

65 Burkholder, W.E. (1969) Cigarette beetle sex attractants. *Cigar Mfr Ass Amer Res Sem:* **19**: 5–6.

66 Burkholder, W.E., Coffelt, J.A., Matsumura, F. and Fonda, H. (1971) Sex attractants for control of the cigarette beetle. *Cigar Mfr Ass Amer Res Sem* **15**: 3–4.

67 Burton, J.S. (1957) Pyrethrum and the Food Trade Part II – Pyrethrum and Synergists. *Food Trade Rev.* 4 pp.

68 Candura, G.S. (1932) Contributo alla conoscenza morfologica e biologica dello struggigrano (*Tenebroides mauritanicus* L.) *Boll Lab Zool Portici* **27**: 1–56.

69 Canzanelli, A. (1935) Contributo alla embriologia e biologia del tarlo del tabacco (*Lasioderma serricorne* Fabricius) *Boll Lab Zool Portici* **27**: 1–56.

70 Carter, S.W. and Dodd, G.D. (1978) The use of ultra low volume pyrethroid sprays for insect control in stored products. *Proc 5th Brit Pest Cont Conf* 4pp.

71 Cavalloro, R. (1966) Effets de l'irradiation gamma sur *Lasioderma serricorne* F (The effects of gamma irradiation on *Lasioderma serricorne* F). *Proceedings of the Fourth International Tobacco Scientific Congress, Athens.*

72 Chadwick, P.R. (1967) Protecting Stored Tobacco from Insects is Essential. *Rhodesian Tob J.* May. 45–47.

73 Chadwick, P.R.L (1986) Surface and Other Factors Modifying the Effectiveness of Pyrethroids Against Insects in Public Health. *Pestic Sci* **16**: 383–391.

74 Chair, M.S., Patel, V.G. and Patel, G.J. (1968) Residual effect of insect growth regulator, methoprene on the development of *Lasioderma serricorne* F. *Ind Soc Tob Sci* Rpt No. 2, 8pp.

75 Chaudrhy and Price (1992) *Pestic Biochem Physiol* **42**: 167.

76 Childs, D.P. (1965) Effect on the cigarette beetle of flowing steam vapor under vacuum. *Tob Sci* **9**: 32–36.
77 Childs, D.P. (1966) Effect on the tobacco moth of flowing steam vapor under vacuum. *Tob Sci* **10**: 113–115.
78 Childs, D.P. and Fletcher, L. (1974) Phosphine (Detia® Gas Ex-B) for Fumigation of Flue-cured Tobacco. *Tob Sci* **18**: 92–94.
79 Childs, D.P. and Overby, J. (1979) Fumigation of Tobacco with Phosphine from Magnesium Phosphide Plates. *J stored Prod Res* **15**: 12–125.
80 Childs, D.P. and Overby, J.E. (1984) Permeability of Coated Fibreboard Panels, Plywood sheets, Polyurethane Foam, and Plastic Coatings to Phosphine. *Tob Sci* **28**: 14–14.
81 Childs, D.P. and Overby, J.E. (1970) Phosphine Fumigation of Tobacco in Louvered Warehouses. *Tob Sci* **14**: 49.
82 Childs, D.P. and Overby, J.E. (1967) Atmospheric chamber fumigation of cigar tobacco against the cigarette beetle. *Tob Sci* **11**: 38–41.
83 Childs, D.P. and Overby, J.E. (1978) Fumigation with ethylene oxide under vacuum: improvement with probe in center of tobacco hogshead. *Tob Sci* **22**: 73–76.
84 Childs, D.P. and Overby, J.E. (1983) Mortality of the cigarette beetle in high CO_2 atmospheres. *J Econ Ent* **76**: 544–546.
85 Childs, D.P. and Tenhet, J.N. (1960) Studies in warehouse fumigation of flue cured tobacco. *Tob Sci* **4**: 42–48.
86 Childs, D.P., Fletcher, L.W., Beard, J.T., Iachetta, F.A. (1983) Cooling tobacco in warehouses during the winter to kill cigarette beetles, Part I: Relevant physical properties of stored tobacco. *Tob Sci* **27**: 116–124.
87 Childs, D.P., Overby, J.E. and Hadden, R.L. (1971a) Phosphine Fumigation of Tobacco in Freight Containers (Part II). *Tob Sci* **15**: 5–6.
88 Childs, D.P., Overby, J.E. and Niffenegger, D. (1971b) Phosphine Fumigation of Tobacco in Freight Containers (Part I). *Tob Sci* **15**: 1–4.
89 Childs, D.P., Overby, J.E. and Nittenegger, D. (1969a) Phosphine Fumigation of Flue-cured Tobacco Warehouse for Control of the Cigarette Beetle. *Tob Sci* **13**: 64–69.
90 Childs, D.P., Overby, J.E. and Watkins, B.J. (1969b) Phosphine Fumigation of Flue-cured Tobacco for Control of the Cigarette Beetle – Laboratory Tests. *Tob Sci* **13**: 49–52.
91 Childs, D.P., Overby, J.E., Cox, E.L. and Niggenegger, D. (1973) Toxicity of Phosphine at Various Concentrations and Temperatures to the Cigarette Beetle. *Agricultural Research Service No. ARS–5–16, USDA/ARS.*
92 Childs, D.P., Overby, J.E., Watkins, B.J. (1968) Low temperature effect on cigarette beetle infestation in tobacco hogsheads. *J Econ Entomol* **61**: 992–996.
93 Chin, W.T. and Lin, H.C. (1979) Studies on the occurrence of cigarette beetle, *Lasioderma serricorne* (F.), with "Serrico" trap in Fen-Yuan tobacco factory. *Bull Taiwan Tob Res Inst* **32**: 53–58.
94 Chuman, T., Kato, K., and Noguchi, M. (1979a) *Agric Biol Chem* **43**: 2005.
95 Chuman, T., Kohno, M., Kato, K. and Noguchi, M. (1979b) 4, 6-demiethyl-7-hydroxy-3-one, sex pheromone of the cigarette beetle (*Lasioderma serricorne* F.) *Tetrahedron letters* **25**, 2361–2364.
96 Chuman, T., Kono, M. and Kato, K. (1982c) United States Patent **4**, 317, 836. March 2.

97 Chuman, T., Mochizuki, K., Kato, K., Ono, M. and Okubo, A. (1983) Serricorone and serricorole, new sex pheromone components of cigarette beetle. *Agric Biol Chem* **47**: 1413–1415.
98 Chuman, T., Mochizuki, K., Mori, M., Kohno, M., Kato, K. and Noguchi, M. (1985) *Lasioderma* chemistry. Sex pheromone of cigarette beetle (*Lasioderma serricorne* F.) *J Chem Ecol* **11**: 417–434.
99 Chuman, T., Mochizuki, K., Mori, M., Kohno, M., Kato, K., Nomi, H. and Mori K. (1982a) Behavioural and electro antennogram responses of male cigarette beetle *Lasioderma serricorne* to optically active serricornins. *Agric Biol Chem* **42**:12.
100 Chuman, T., Mochizuki, K., Mori, M., Kohno, M., Ono, M., Ohnishi, I. and Kato, K. (1982b) The pheromone activity of (±)Serricornins for male cigarette beetle (*Lasioderma serricorne* F.). *Agric Biol Chem* **46**: 593–595.
101 Cline, L.D. and Highland, H.A. (1985) Survival, reproduction and development of seven species of stored-product insects on the various food components of lightweight, high-density, prototype military rations. *J Econ Entomol* **78**: 779–782.
102 Coffelt, J.A. (1975) Multiple mating by *Lasioderma serricorne* (F.) – Effects on fertility and fecundity. *Proc 1st Int Wksp S-P Prot* Savannah, Ga.
103 Coffelt, J.A. and Burkholder, W.E. (1972) Reproductive biology of the cigarette beetle, *Lasioderma serricorne* 1. Quantitative laboratory bioassay of the female sex pheromone from females of different ages. *Ann Entomol Soc Am* **65**: 447–450.
104 Coffelt, J.A. and Vick, K.W. (1973) A black mutation of *Lasioderma serricorne* (F.)(Coleoptera: Anobiidae). *J stored Prod Res* **9**: 65–70.
105 Cooper, L.M. and Bengston, M. (1974) Tobacco beetle and its control. *Aust Tob Grow Bull* **21**: 36–40.
106 Corbet, A.S. and Tams, W.H.T. (1943) Keys for identification of the Lepidoptera infesting stored products. *Proc Zool Soc Lond* **113**: 55–148.
107 C.O.R.E.S.T.A. (1975) *Ephestia elutella, Lasioderma serricorne* (Presence dans le monde, dégats et moyens de lutte).
108 C.O.R.E.S.T.A. (1989) Analyses of methoprene, GC and HPLC *C:O:R:E:S:T:A: Subgroup on pesticide residues*, doc.93/19.
109 C.O.R.E.S.T.A. (1993) The use of published guidelines in the selection, application, and handling of agrochemicals in leaf tobacco production. *C.O.R.E.S.T.A. Bull* **1993–1**: 21–31.
110 Cotton, R.T. (1923) *Aplastamorpha vandinci* Tucker, an important parasite of *Sitophylus oryza* L *J Agric Res* **23**: 549–556.
111 Cox, P.D. and Bell, C.H. (1981) A review of the biology of moth pests on stored products. *ADAS* publication, Slough Laboratory, Berks.
112 Davis, R. (1983) A Review of In-transit Shipboard Fumigation of Grain-Methodology, Efficacy and Safety. *3rd International Working Conference on Stored-product Entomology*. Manhattan, KS, USA, October 23–28, pp. 347–356.
113 Desmarchelier, J.M. (1988) The relationship between wet-bulb temperature and the intrinsic rate of increase of eight species of stored-product Coleoptera. *J stored Prod Res* **24**: 107–113.
114 Dévai, I., Felföldy, L., Wittner, I and Plósz, S. (1988) Detection of Phosphine: New Aspects of the Phosphorous Cycle in the Hydrosphere. *Nature* **333**: 343–345.
115 Dickins, G.R. (1936) The scent glands of certain Phycitidae (Lepidoptera). *Trans R Entomol Soc Lond* **85**: 331–362.

116 Dillon, E.S. and Dillon, L.S. (1972) *A manual of common beetles of eastern, North America*. Vols. 1 and 2. Dover publications, New York, 894 pp.

117 Dodd, G.D. (1983) Space Spraying in Practice – A Review. *6th British Pest Control Association Conference* (Cambridge) 16 pages.

118 Dowd, P.F. (1989) *In situ* production of hydrolytic detoxifying enzymes by symbiotic yeasts in the cigarette beetle (Coleoptera, Anobiidae). *J Econ Entomol* **82**: 396–400.

119 Dowd, P.F. and Shen, S.K. (1990) The contribution of symbiotic yeast to toxin resistance of the cigarette beetle (*Lasioderma serricorne*). *Entomol exp appl* **56**: 241–248.

120 Ducom, P. and Bourges, C (1986) Determination of Phosphine Concentrations with an Electrochemical Cell. *Proceedings of the 4th International Working Conference on Stored-product Protection*, Tel Aviv, Israel, September, pp. 630–635.

121 Edmond, D.E. (1971) Preliminary Investigations of the Ability of Phosphine Gas to Penetrate Polyethylene Tobacco Case Liners. *Tob Sci* **15**: 11.

122 Edmond, D.E., Hadden, R. and Yu, G. (1971) The Penetration of Phosphine Gas into Lined Tobacco Cases During Atmospheric Fumigation. *Tob Sci* **15**: 80–83

123 Edwards, J.P., Short, J.E. and Rowlands, D.G. (1988) Metabolism and insect juvenile hormone activity of methoprene and three major metabolites in *Tribolium casteneum*, (Herbst.) pupae. *J stored Prod Res* **24**: 165–172.

124 Edwards, S.R., Bell, B.M. and King, M.E. (eds) (1980) Pest control in museums: a status report. *The Association of Systematics Collection*, U.S.A.

125 Ehley, K.M. (1991) *Introduction to the concepts of variation*. Creative Quality Solutions Inc. Wauwatosa, USA. video.

126 Elliot, M., Janes, N.F. and Potter, C. (1978) The Future of Pyrethroids in Insect Control. *Ann Rev Entomol* **23**: 443–469.

127 Els, J.M., de Viliers, W.M. and Matthee, J.J. (1978b) The effect of laboratory rearing medium on the sensitivity of *Lasioderma serricorne* (F.) for two residual insecticides. *Phytophylactica* **10**: 129–130.

128 Els, J.M., Thomas, A.C., de Villiers, W.M., Matthee, J.J. (1978a) Lethal effect of gamma radiation on *Lasioderma serricorne*. *Phytophylactica* **10**: 127–128.

129 Els, J.M., Wik, R.J.V. and Wolmarans, E. (1978c) Residual efficacy of tetrachlorvinphos against *Lasioderma serricorne*. *Phytophylact* **10**: 81–84.

130 Emden, F. van (1931) Zur kenntnis der morphologie und okologie des brotk_fer parasiten, *Cephalonomia quadridentata* Duchaussoy. *Zeit Morph Okol Tiere* **23**: 425–574.

131 Faustini, D.L. (1983) Magnesium Phosphide as a Fumigant for the Control of the Cigarette Beetle, *Lasioderma serricorne* (F.) at Low Temperature. *Proceedings of the 3rd International Working Conference on Stored-product Entomology*. Kansas State University, Manhattan, KS, USA, October 23–28, pp. 308–328.

132 Faustini, D.L. (1985) Stored insect control: A biotechnical approach to cigarette beetle monitoring. *Tob Rep* **112**: 44–47.

133 Faustini, D.L., Barak, A.V., Burkholder, W.E. and Leos-Martinez, J. (1991) Combination-type trapping for monitoring stored-product insects – a review. *J Kansas Ent Soc* **63**: 539–547.

134 Faustini, D.L., Lehman, R.M., Long, J.S. and Manzelli, M.A. (1987) The efficacy of Kabat as a surface spray on oriental tobacco bales for control of the cigarette beetle, *Lasioderma serricorne* (F.) (Coleoptera: Anobiidae). *Beit Tabak Int* **14**: 41–44.

135 Fletcher, L.W. (1977) A procedure for collecting large numbers of eggs of *Lasioderma serricorne* (F.). *J stored Prod Res* **13**: 87–88.

136 Fletcher, L.W. (1980) Rearing the cigarette beetle on tobacco in the laboratory. *Tob Sci* **24**:59–60.

137 Fletcher, L.W. and Childs, D.P. (1976) Plastic Sheets for Protecting Stored Tobacco from the Cigarette Beetle. *Tob Sci* **20**: 14–16.

138 Fletcher, L.W. and Long J.S. (1973) Evaluation of an electric grid light trap as a means of sampling populations of the cigarette beetle. *Tob Sci* **17**: 37–39.

139 Fletcher, L.W. and Long, J.S. (1971) Influence of food odours on oviposition by the cigarette beetle on nonfood materials. *J Econ Entomol* **64**: 770–771.

140 Fletcher, L.W. and Long, J.S. (1976) Cigarette beetle oviposition and egg hatch at selected temperatures. *Tob Sci* **20**: 49–50.

141 Fletcher, L.W., Childs, D.P. (1976) Plastic sheets for protecting stored tobacco from the cigarette beetle. *Tob Sci* **20**: 14–16.

142 Fletcher, L.W., Knülle, W., Childs, D.P., Spadafora, R.R., Long, J.S. and Delamar, C.D. (1973) Responses of cigarette beetle larvae to temperature and humidity. *ARS–S–22*, Agricultural Research Service, U.S. Department of Agriculture.

143 Fletcher, L.W., Long, J.S., Delamar, C.D. (1973b) Low temperature and infestations of beetles in tobacco hogsheads stored in field warehouses. *Tob Sci* **17**: 163–164.

144 Fletcher, L.W., Overby, J.E., Childs, D.P., Dean, T.A. (1974) Redrying of tobacco strips and its effect on mortality of cigarette beetles. *Tob Sci* **18**: 35–36.

145 Fluck, E. (1973) The Chemistry of Phosphine. *Fortschritte Chemischen Forschung* **35**: 1–64,.

146 Fraenkel, G. and Blewitt, M. (1943) The natural foods and food requirements of several species of stored product insects. *Trans R Ent Soc* **93**: 457–490.

147 Fraenkel, G. and Blewitt, M. (1946) The dietics of the clothes moth *Tineola bisselliella* Hum. *J Exp Biol* **22**: 156–161.

148 Fraenkel, G. and Blewitt, M. (1946a) The dietics of the caterpillars of three *Ephestia* species, *E. kuehniella, E. elutella* and *E. cautella*, and of a closely related species, *Plodia interpunctella*. *J Exp Biol* **22**: 162–171.

149 Frank, R. and Ripper, G. (1987) Fate of Phosphine in the atmosphere. Report of Blattelle Inst. to Detia Degesch, January, 11pp.

150 Friemel, W. (1984) The "Detia Bag Blanket", Its Use and Application. pp. 359–367. In *Controlled Atmospheres and Fumigation in Grain Storages* (Ripp, B. E.-Ed.). Elsevier, Oxford, UK. 798 pp.

151 Frisch, A.F., Francis, B.W., Chavis, M.K. and Jenkins, R.W. (1982) Cigarette smoke formation studies VII. The fate of (5-^{14}C)-methoprene in cigarette smoke. *Beit Tabak Int* **11**: 203–208.

152 Frost, S.W. (1952) Light traps for insect collection, survey and control. *The Pennsylvania State College School of Agriculture, Agricultural Experiment Station, State College, Pennsylvania Bulletin* 550, April 1952.

153 Garry, V.F. Griffith, J., Danzl, T.J., Nelson, R.Ll, Whorton, E., Kreuger, L.A. and Cervenka, J. (1989) Human Genetoxicity: Pesticide Application and Phosphine. *Science* **246**: 251–255.

154 Geneve, R., Delon, R. and Cocaign, J.M. (1987) Comparison of different trap systems as means of sampling populations of *Lasioderma serricorne* in tobacco warehouses. *Annales du Tabac Seita Paris France* **21**, Section 2, 99–108.
155 Geneve, R., Delon, R., Fleurat-Lessard, F., Hicaube, D. and LeTorc'h (1990) Comparison of different types of pheromone traps for monitoring the presence of *Ephestia elutella* (Hubner) (Lep. Pyralidae) in tobacco warehouses. *Ann du Tabac* **19**: 65–77.
156 Geneve, R., Ducom, P., Vanbranteghem, V. and Delon, R. (1985). Desinsectisation d'un entrepot de tabac de 200'000 mètres carrés par le phosphure d'hydrogene. (Desinfestation of a tobacco warehouse by phosphine). *Proc C.O.R.E.S.T.A.*, Paris, (1985), 10pp.
157 Gerard, D., Kraus, J., Quirin, K.-W. (1988a) Rückstandsfreie Druckentwesung mit natürlicher Quellkohlensäure. *Gordian* **88**: 90–93.
158 Gerard, D., Kraus, J., Quirin, K.-W., Wohlgemuth, R. (1988b) Anwendung von Kohlendioxid (CO_2) unter Druck zur Bekämpfung vorratsschädlicher Insekten und Milben (Use of pressurized carbon dioxide (CO_2) to combat pest insects and mites). *Pharm Ind* **50**: 1298–1300.
159 Goodwin-Bailey, K.F. (1960) Synergism of Piperonyl Butoxide with Pyrethrins against Several Species of Insect. *Chemistry and Industry*, June, 700–702.
160 Green, A.A., Kane, J. and Gradidge, J.M.G. (1966) Experiments on the control of *Ephestia elutella* (HB) (Lepidoptera: Phycitidae) using dichlorvos vapour. *J stored Prod Res* **2**: 147–157.
161 Halstead, D.G.H. (1963) External sex differences in stored-product Coleoptera. *Bull Ent Res* **54**: 119–134.
162 Halstead, D.G.H. (1986) Keys for the identification of beetles associated with stored products. I Introduction and key to families. *J stored Prod Res* **22**: 163–203.
163 Heckman, R.A. and Conner, T.R. (1990) Solid-phase Extraction of Methoprene Residue From Tobacco With On-line Elution and HPLC Analysis. LC*GC **10**: 855–856.
164 Heckman, R.A., Ferguson, B.S., Fan, T.S., Mei, J.V., Yin, C.-M., Conner, T.R., Addington, V.W. and Benezet, H.J. (1992) Validation of an enzyme immunoassay for analysis of methoprene residues on tobacco. *J Agr Food Chem* **40**: 2530–2532.
165 Heeman, V. (1986) Insect Control: Natural chemicals attack beetles. *Tob Rep* **113**, 30–40.
166 Henrick, C.A., Anderson, R.J., Staal, G.B. and Ludvik, G.F. (1978) Insect juvenile hormone activity of optically active alkyl (2E,4E)–3,7,11-trimethyl-2,4-dodecadienoates and of arylterpenoid analogues. *J Agr Food Chem* **26**: 542–550.
167 Henrick, C.A., Willy, W.E., Garcia, B.A. and Staal, G.B. (1975) Insect juvenile hormone activity of the stereoisomers of ethyl 3,7,11-trimethyl-2,4-dodecadienoate. *J Agr Food Chem* **23**: 396–400.
168 Highland, H.A. (1988). Vacuumized and unvacuumized polyethylene or polyester film bags: insect survival and resistance to insect penetration. *J Econ Ent* **81**: 955–958.
169 Highland, H.A., Guy, R.H. (1969) Insect resistance of polypropylene film overwrap on cigarette packages. *Tob Sci* **13**: 121–122.
170 Highland, H.A., Schoenherr, W.H., Winburn, T.F. and Lawson, D.E. (1979)

Phosphine and methyl bromide fumigation of commodities in woven plastic or paper bags. *Cereal Foods World* **24**: 19–22.

171 Hirose, T., Abe, I., Kohno, M., Suzuri, T., Oshima, K., Okakura, T. (1975) The use of microwave heating to control insects in cigarette manufacture. *J Microwave Power* **10**: 181–190.

172 Hole, B.O., Bell, C.H., Mills, K.A. and Goodship, G. (1976) The Toxicity of Phosphine to all Developmental Stages of Thirteen Species of Stored Product Beetles. *J stored Prod Res* **12**: 235–244.

173 Howe, R.W. (1957) A laboratory study of the cigarette beetle, *Lasioderma serricorne* (F.)(Col., Anobiidae) with a critical review of the literature on its biology. *Bull Ent Res* **48**:9–56.

174 Howe, R.W. (1965) A summary of estimates of optimal and minimal conditions for population increase of some stored product insects. *J stored Prod Res* **1**: 177–184.

175 Howe, R.W. (1974) Problems in the Laboratory Investigation of the Toxicity of Phosphine to Stored Product Insects. *J stored Prod Res* **10**: 167–181.

176 Hutchens, H.M. (1990) Advantages of Using Insect Growth Regulators. *Federation of Asian and Oceania Pest Managers Association Convention*, Hong Kong, October 17, 1990.

177 Imai, T., Kodama, H., Chuman, T. and Kohno, M. (1990) Female produced oviposition deterrents of the cigarette beetle, *Lasioderma serricorne (F.)* (Coleoptera: Anobiidae). *J Chem Ecol* **16**: 1237–1247.

178 I.S.O. (1991) *ISO 9000, International standards for quality management*. I.S.O., Geneva.

179 Jacob, S.A. and Mohan, M.S. (1973) Predation on certain stored product insects by red flour beetle. *Indian J Entomol* **35**: 95–98.

180 Jay, E., Davis, R. and Zehner, J.M. (1983) In-transit Fumigation of Truck-ship Containers with Hydrogen Phosphide. *U.S.D.A.. Agricultural Research Service* No. AAT-5–28, 13 pp.

181 Jay, E.G. (1980) Suggested Conditions and Procedures for Using Carbon Dioxide to Control Insects in Grain Storage Facilities. *U.S.D.A., ARS Bulletin* 51–26 6pp. Sept.

182 Jepson, W.F. (1939) Entomology division reports. *Rep Dept Agric Mauritius* (1937) pp. 41–46 and (1938) pp. 40–51.

183 Jones, C.R. (1913) The cigarette beetle (*Lasioderma serricorne* Fabr.) in the Philippine Islands. *Philippine J Sci D. Gen Biol, Ethnol Anthropol.* **8**: 1–61.

184 Jones, O.T. (1991) Monitoring and control of insect pests with pheromones and other semiochemicals. *Agro-ind Hi-tech* **2**: 27–32.

185 Jurzitza, G. (1969) The development of normal and aposymbiotic larvae in tobacco with differing nicotine content *Z ang Ent* **63**:233–236.

186 Jurzitza, G. (1972) The function of the endosymbiotic algae of the cigarette beetle *Lasioderma serricorne* F. (Col., Anobiidae) when reared under conditions of protein deficiency. *Z ang Ent* **72**: 198–205.

187 Jurzitza, G. (1977) Electron microscopic studies on the symbiont inhabited ceca of the larval midgut of *Lasioderma serricorne* F. (Coleoptera: Anobiidae) *Z. Parasit* **54**: 193–207.

188 Kaelin, P., Morel, P. and Gadani, F. Isolation of *Bacillus thuringiensis* from stored tobacco and *Lasioderma serricorne* (F.). *Appl Env Microbiol* **60**: 19–25.

189 Kashi, K.P. (1982) Dose-mortality Responses of Five Species of Stored–products Insects to Phosphine. *Int Pest Control* **24**: 46–49.

190 Kashi, K.P., Mutho, M. and Majumder, S.K. (1977). Rapid evaluation of phosphine permeability through various films and coated fabrics. *Pestic Sci* **8**: 492–496.

191 Kearns, C.W. (1934) A hymenopterous parasite (*Cephalonomia gallicola* Ashm.) new to the cigarette beetle (*Lasioderma serricorne* Fab). *J Econ Ent* **27**: 801–806.

192 Keever, D.W. (1989) Use of carbon dioxide to disinfest a tobacco warehouse of the cigarette beetle. *J Agric Ent* **6**: 43–51.

193 Keever, D.W. (1993) Freezing of tobacco display samples for disinfestation of the cigarette beetle (Coleoptera: Anobiidae). *Tob Sci* **37**: 65–68.

194 Kellen, W.R. and Hoffmann, D.F. (1987) Effect of two nonocluded viruses of *Amyelois transitella* on larvae of *Cadra figulilella* and *Ephestia elutella* (Lepidoptera: Pyralidae) in Laboratory tests. *Env Ent* **16**: 630–632.

195 Keuchenius, P.E. (1917) Waarnemingen over ziekten en plagen bij tabak (derde serie). De tabaks boeboek (*Lasioderma*) en de tabaksmot (*Setomorpha*). *Meded besoek Proefst* **26**: 56 pp.

196 Khan, M.A. (1983) Effect of relative humidity on adults of 10 different species of stored product beetles. *Z ang Ent* **95**: 217–227.

197 Khan, M.A. (1983b) Invasion von Vorratsschädlingen durch Verschlüsse (Invasion of stored product insects through seals). *Anz Sch_dlingskde, Pflanzenschutz, Umweltschutz* **56**: 91–94.

198 Khan, M.A. (1983a) Untersuchungen über die Invasion von Eilarven von vorratsschädlichen Insekten durch verschieden grosse Poren des Verpackungsmaterials (Investigation on the invasion of egg-larvae of stored product insects through different sizes of pores of the packaging material). *Anz Schädlingskde, Pflanzenschutz, Umweltschutz* **56**: 65–67.

199 Kirkaicharian, S.B. (1974) Tobacco Fumigation with Phosphine. *Tob Int*, pp. 38–39, 13 September.

200 Klein, R.M. (1986) Pictorial guide for rapid identification of common adult storage insects. *J Food Protection* **49**: 154–160.

201 Kohno, M. (1982) A rapid suction method for selective collection of small beetles from mass culture. *Appl Ent Zool* **17**: 135–136.

202 Kohno, M. and Ohnishi, A. (1986) Some factors affecting the oviposition behaviour of cigarette beetle (Coleoptera, Anobiidae) *Kontyû, Tokyo* **54**: 29–32.

203 Kohno, M., Chuman, T., Kato, K. and Noguchi, M. (1983) The olfactory response of the cigarette beetle, *Lasioderma serricorne* Fabricius, to various host foods and cured tobacco extracts. *Appl Ent Zool* **18**: 401–406.

204 Kohno, M., Mochizuki, K., Chuman, T. and Ohnishi, A. (1986) Pheromone-like substances affecting the oviposition behaviour of the female cigarette beetle, *Lasioderma serricorne* (F.) (Coleoptera: Anobiidae). *Appl Ent Zool* **21**: 15–20.

205 Koizumi, S., Kikukawa, H. and Shimojic, C. (1985) Occurrence and ecology of pests attacking stored tobacco in growing districts in Kyushu. *Kagoshima Tob Exp Stm Bull* **26**: 75–86.

206 Krasnoff, S.B. and Vick, K.W. (1984) Male wing-gland pheromone of *Ephestia elutella*. *J Chem Ecol* **10**: 667–679.

207 Krasnoff, S.B., Vick, K.W. and Mankin, R.W. (1983) Female calling behaviour in *Ephestia elutella* and *E. figulilella* (Lepidoptera: Pyralidae) *Florida Ent* **66**: 249–254.

208 Kunjeku. E. (1986) Comparative efficacy and persistence of four synthetic pyrethroids on different surfaces for control of tobacco beetle, *Lasioderma serricorne* (F). *Proc. CORESTA Symp.* Taormina, Italy, (1986), 350pp.

209 Kurup, A.R. (1961) Preliminary studies on the nutritional response of the cigarette beetle (*Lasioderma serricorne* F.) to different grades of Flue-cured tobacco. *Indian Tob J* **11**: 91–95.

210 Kurup, A and Parkhe, D.P. (1961) Some observations on the biology of the cigarette beetle (*Lasioderma serricorne* F.) *Ind J Ent* **23**: 274–278.

211 LeCato, G.L. (1978) Functional response of red flour beetles to density of cigarette beetles and the role of predation in population regulation. *Environ Ent* **7**: 77–80.

212 Leesch, J.G. (1982) Accuracy of Different Sampling Pumps and Detector Tube Combinations to Determine Phosphine Concentrations. *J Econ Ent* **75**: 899–905.

213 Leesch, J.G. and Highland, H. A. (1978) Fumigation of Shrink-wrapped Pallets. *J Georgia Ent Soc* **13**: 43–50.

214 Leesch, J.G. and Highland, H.A. (1982) Phosphine and Methyl Bromide Permeation of Fabric and Multiwall Paper Bags. *J Econ Ent* **75**: 819–821.

215 Lefkovitch, L.P. (1963) Census studies on unrestricted populations of *Lasioderma serricorne* (F.) (Coleoptera: Anobiidae). *J Anim Ecol* **32**: 221–231.

216 Lefkovitch, L.P. (1964) The growth of restricted populations of *Lasioderma serricorne* (F.) (Coleoptera, Anobiidae) *Bull Ent Res* **55**: 87–96.

217 Lefkovitch, L.P. (1967) Larval movement of *Lasioderma serricorne* (F.) (Coleoptera: Anobiidae) in a plane. *Res Pop Ecol* **9**: 45–52.

218 Lefkovitch, L.P. and Currie, J.E. (1963) The effects of food shortage upon larvae of *Lasioderma serricorne* (F.) (Coleoptera, Anobiidae). *Bull Ent Res* **54**: 535–547.

219 Leidy, R.B., Sheets, T.J. and Wright, C.G. (1987). DDVP residues in air following application to a tobacco-storage warehouse. *Env Monit Assess* **9**: 263–268.

220 Lent, H. and Deoliviera, S.J. (1968) Spider *Uloborus geniculatus* (Oliv.). Predator of the cigarette beetle *Lasioderma serricorne*, F. and the possibility of its use in biological control. *Rev Brazil Biol* **24**: 461–463.

221 Lever, R.J.A.W. (1941) Entomological Notes. *Agric J Fijii* **12**: 19–22.

222 Levinson, A.R. and Buchelos, C.Th. (1988) Population dynamics of *Lasioderma serricorne* F. (Col., Anobiidae) in tobacco stores with and without insecticidal treatments: a three year survey by pheromone and unbaited traps. *J Appl Ent* **106**: 201–211.

223 Levinson, A.R. and Levinson, H.Z. (1986a) Restrained pheromone responses of male tobacco beetles (*Lasioderma serricorne* F.) to 2S,3S- anhydroserricornin in presence of 4S,6S,7R-serricornin. *J Appl Ent* **101**: 282–287.

224 Levinson, A.R. and Levinson, H.Z. (1986b) Antagonised pheromone responses of male tobacco beetles (*Lasioderma serricorne* F.) to erythro-diasttereoisomers of 4S,6S,7S–serricornin. *Naturwissenschaften* **73**: 36–37.

225 Levinson, H.Z, Levinson, A-R., Franke, W. and Heemann, V. (1983) UK Patent Application GB 2104061 A.

226 Levinson, H.Z. and Levinson, A.R. (1987) Pheromone biology of the tobacco beetle (*Lasioderma serricorne* F., Anobiidae) with notes on the pheromone antagonism between 4S,6S,7S-serricornin and 4S,6S,7R-serricornin. *J Appl Ent* **103**: 217–240.

227 Levinson, H.Z., Levinson, A.R., Francke, W., Mackenroth, W. and Heeman, V. (1981) The pheromone activity of anhydroserricornin and serricornin for male cigarette beetles (*Lasioderma serricorne* F.) *Naturwissenschaften* **68**: 148–149.

228 Levinson, H.Z., Levinson, A.R., Kahn, G.E. and Schafer, K. (1983) Occurrence of a pheromone producing gland in female tobacco beetles. *Experentia* **39**: 1095–1097.

229 Lindgren, D.L. and Vincent, L.E. (1966) Relative Toxicity of Hydrogen Phosphide to Various Stored–product Insects. *J stored Prod Res* **2**: 141–146.

230 Livingstone, E.M. and Reed, W.D. (1936) Insect fauna of cured tobacco in storage in the United States. *J Econ Ent* **29**: 1017–1022.

231 Long, J.S., Lehman, R.M. and Manzelli, M.A. (1978) Altosid, an examination of its activity against the cigarette beetle. *Tob Sci* **22**: 151–152.

232 Long, J.S., Lehman, R.M., Tickle, M.H. and Manzelli, M.A. (1980) The Efficacy of Kabat for Control of the Cigarette Beetle Under Severe Infestation Conditions. *Tob Sci* **24**: 119–121.

233 Manzelli, M.A. (1987) Control of the cigarette beetle in stored and processed tobacco. *Tob J Int* **6/87**: 373–378.

234 Manzelli, M.A. (1982) Management of Stored Tobacco Pests, The Cigarette Beetle (Coleoptera: Anobidae) and Tobacco Moth (Lepodoptera: Pyralidae), with Methoprene. *J Eco Ent* **75**: 721–723.

235 Manzelli, M.A. and Lehman, R.M. (1977) Altosid: An Insect Growth Regulator for the Control of the Cigarette Beetle in Stored Tobacco. *27th Tobacco Workers Conference*, Atlanta, Georgia, January 10–13.

236 Marshall, T.A. (1888) Braconidae, in André, Species des Hyménoptères d'Europe et d'Algérie 4. Beaune, *in* Richards and Thompson, (1932).

237 Marzke, F.O., Coffelt, J.A. and Silhacek, D.L (1977) Impairment of reproduction of the cigarette beetle, *Lasioderma serricorne* (Coleoptera: Anobiidae) with the insect growth regulator, methoprene. *Ent Exp Appl* **22**: 294–300.

238 Matthews, G.A. (1992) *Pesticide application methods*. Longman and Wiley. New York. 2nd ed. 405pp.

239 McLellan, R.H. (1962) Practical trials of fogging tobacco warehouses with synergised pyrethrum in oil solutions to control the adult cigarette beetle. *African Pyrethrin* **1962**: 26–28.

240 McClelian, W.G. and Yu, G. (1973) Retention of Phosphine Gas with 3ml. Polyethylene. *Tob Sci* **17**: 24.

241 Meyer, A. (1980) Pest control in stored tobacco. *Zimbabwe Sci News* **14**: 101–103.

242 Milne, D.L. (1963) A study of the nutrition of the cigarette beetle, *Lasioderma serricorne* F., (Coleoptera: Anobiidae) and a suggested new method for its control. *J Ent Soc South Africa* **26**: 43–63.

243 Minor, M.F. (1979) Do adult cigarette beetles feed? *Tob Sci* **23**: 61–64.

244 Minor, M.F., Manzelli, M.A., Lehman, R.M. and Long, J.S. (1983) A large-scale evaluation of Kabat for the control of *Lasioderma serricorne* (F). *Tob Sci* **27**: 64–65.

245 Mochizuki, K., Chuman, T., Mori, M. Kohno, M. and Kato, K. (1984) Activity of the stereoisomers of serricornin, sex pheromone of the cigarette beetle (*Lasioderma serricorne* F). *Agric Biol Chem* **48**: 2833–2834.

246 Mochizuki, K., Mori, M., Chuman, T., Kohno, M., Ohnishi, A., Watarabe, H.

and Mori, K. (1986) Reinvestigation of anhydroserricornin in (2S, 3S)-2-6-diethyl-3,5, dimethyl-3,4-dihydro-2H-pyran, as a sex pheromone component for male cigarette beetle. *J Chem Ecol* **12**: 179–186.

247 Mokrzhetskii, S.A. and Bragina, A.T. (1915) Report of the entomological laboratory of the experimental station of Salgir for 1913–14. Simferopol. Crimea. 9pp. (Russian) Abstract in *Rev Appl Ent* (A) **3**: 613.

248 Montazer, A. di M.-F. de C.-R. (1975) Esperienze Sulla Lotta Control il Tarlo del Tabacco a Mezzo della Fosfina. *Informatore Fitopatologico* **3**: 65–74.

249 Morgan, A.C. (1913) An enemy of the cigarette beetle. *Proc Ent Soc Wash* **15**: 89.

250 Mori, M., Mochizuki, K., Kohno, M., Chuman, T., Ohnishi, A., Watarabe, H. and Mori, K. (1986) Inhibitory action of (4S, 6S, 7S)-isomer to the pheromonal activity of serricornin, 4S, 6S, 7S -7-hydroxy-4,6-dimethyl-3-nonane. *J Chem Ecol.* **12**: 83–89.

251 Mossop, M.C. (1932) Pests of stored tobacco in Southern Rhodesia. *Rhod Agric J* **29**: 245–265.

252 Mueller, D., Pierce, L. Benezet, H. and Krischik, V. (1990) Practical application of pheromone traps in the food and tobacco industry. *J Kansas Ent Soc* **63**: 548–553.

253 Muenzel, M. (1987) Examination of efficacy of magnesium phosphide (Detiaphos R) against the cigarette beetle *Lasioderma serricorne* (Fabr.) (Col., Anobiidae) at low temperatures. *Anz Schad, Pflan, Umwelt* **60**: 74–78.

254 Mullen, M.A. and Arbogast, R.T. (1979) Time-Temperature-Mortality relationships for various stored product insect eggs and chilling times for selected commodities. *J Econ Entomol* **72**: 476–478.

255 Mullen, M.A. and Highland, H.A. (1990) Development of six species of stored product insects on Multi-grain blend supplement. *J Entomol Sci* **25**, 1: 21–24.

256 Müller-Mohnssen, H. (1991) Insecticides: science ineffective as an early warning system. *German Med J* **88**: C, 1966–1971.

257 Munro, J.W. and Thomson, W.S. (1929) Report on insect infestation of stored cacao. *EMB*, London, **24** (in Richards and Thomson, 1932).

258 Navarro, S., Calderon, M. (1972) Exposure of *Ephestia elutella* (Lepidoptera, Phycitidae) to low pressures: Effects on adults. *J stored Prod Res* **8**: 209–212.

259 Nemoto, T., Kuwahara, Y. and Suzuki, T. (1987) Interspecific differences in *Venturia* kairomones in larval faeces of four stored phycitid moths. *Appl Ent Zool* **22**: 553–559.

260 Niiho, C. (1984) Ecological study of the tobacco beetle, *Lasioderma serricorne* (F.) II Growth of tobacco beetle fed on breadcrumbs. *Jap J appl Ent Zool* **28**: 209–216 (Japanese).

261 N.I.O.S.H. (1990) National Institute for Occupational Safety and Health, *Guide to Chemical Hazards*. Publication No. 90–117.

262 Oh, M.H. (1984) Effects of tobacco manufacturing process on the mortality of cigarette beetle, *Lasioderma serricorne* F. (Coleoptera, Anobiidae). *J Korean Soc Tob Sci.* **6**: 94–96.

263 Olkowski, H. and Olkowski, W. (1976) Entomophobia in the urban ecosystem; some observations and suggestions. *Bull Ent Soc Am* **22**: 313–317.

264 Olsen, A.R., Bryce, J.R., Lara, J.R., Madenjian, J.J., Potter, R.W., Reynolds, G.M. and Zimmerman, M.L. (1987) Survey of stored product and other economic pests in import warehouses in Los Angeles. *J Econ Ent* **80**: 455–459.

265 Pant, N.C. (1972) Physiology of symbiotes in insects. *Insect Physiology and Anatomy* Eds Pant, N.C. and Ghai, S. Commonwealth Inst. Entomol., London.

266 Pant, J.C. and Anand, M. (1985) Qualitative vitamin requirement of *Lasioderma serricorne* (Fab.) larvae. *J Ent Res* **9**: 165–169.

267 Pant, J.C. and Fraenkel, G. (1950) The function of the symbiotic yeasts of two insect species, *Lasioderma serricorne* F. and *Stegobium* (Sitrodepa) *paniceum* L. *Science* **112**: 498–500.

268 Patel, G.J. Chari, M.S. and Bharpoda, T.M. (1986). Protection of Bidi tobacco seed from cigarette beetle (*Lasioderma serricorne* FB) infestation in storage. *Ind. Tob. J.* **17**: 3–5.

269 Patourel, G.N.J.Le and Zhou, J.J. (1990) Action of amorphous silica dust on the german cockroach, *Blattella germanica* (Lineaus) (Orthoptera: Blattidea). *Bull Ent Res* **80**: 11–17.

270 Pendlebury, J.B., Jefferies, D.J., Banham, E.J., Bull, J.O. (1966) Some effects of gamma radiation on *Rhizopertha dominica* (F), *Cadra cautella* (WLK), *Plodia interpunctella* (Hübn.) and *Lasioderma serricorne* (F). In: Connwell, P.B. (ed.): *The Entomology of radiation disinfestation of grain*. Pergamon Press.

271 Phelan, P.L. and Baker, T.C. (1986) Male-size-related courtship success and intersexual selection in the tobacco moth, *Ephestia elutella*. *Experientia* **42**: 1291–1293.

272 Phelan, P.L., Silk, P.J., Northcott, C.J., Tan, S.H. and Baker, T.C. (1986) Chemical identification and behavioural characterization of male wing pheromone of *Ephestia elutella* (Pyralidae) *J Chem Ecol* **12**: 135–146.

273 Porta, A. (1929) *Fauna Coleopterum Italica*. Vol III. Diversicornia 466 pp. Piacenza.

274 Potter, C. (1935) An account of the constitution and use of an atomised white oil-pyrethrum fluid – to control *Plodia interpunctella* and *Ephestia elutella* in warehouses. *Ann appl Biol* **22**: 769–805.

275 Powell, T.E. (1931) An Ecological Study of the Tobacco Beetle, *Lasioderma serricorne* Fabr., with special reference to its life history and control. *Ecol Monogr* **1**:333–393.

276 Press, A.F. and Childs, D.P. (1966) Control of the tobacco moth with dichlorvos. *J Econ Ent* **59**: 264–265.

277 Price, N.R. (1984) Active exclusion of phosphine as a mechanism of resistance in *Rhyzopertha dominica* (F) (Coleoptera: Bostrychidae). *J stored Prod Res* **20**: 163–168.

278 Quistad, G.B., Staiger, L.E. and Schooley, D.A. (1975) Environmental degradation of the insect growth regulator methoprene (Isopropyl (2E,4E)-11-methoxy-3,7,11-trimethyl-2,4-dodecadienoate). III. Photodecomposition. *J Agr Food Chem* **23**: 299–303.

279 Rabindra, R.J., Jayaraj, S. and Balasubramanian, M. (1987) Occurrence of eugregarine, *Pyxinia* sp. in the larvae of the cigarette beetle, *Lasioderma serricorne* Fb. *J Ent Res* **11**: 207–208.

280 Rangaswamy, J.R. (1985) Sex pheromones of stored product insects. *J Sci Ind Res* **44**: 491–499.

281 Rao, R.S.N. (1977) New record of mites *Cheyletus eruditus* (Schrank) as a predator on eggs of *Lasioderma serricorne* F. *Tob Res* **3**: 120–122.

282 Rassmann, W. (1986) Investigations on the efficiency of *Bacillus thuringiensis* for the control of moths in stored grain. *Mach Devt Pflan* **38**: 61–63.

283 Rassmann, W. (1980) Zur Bekämpfung von Larven des Tabakkäfers (*Lasioderma serricorne* Fab.) durch Gefrieren von Rohtabak und Tabakfertigprodukten (Control of larvae of the cigarette beetle (*Lasioderma serricorne* Fab.) by freezing tobacco and tobacco products). *Anz Schädlingskde, Pflanzenschutz, Umweltschutz* **53**: 133–135.

284 Redlich, H., Samm, K., Lenfers, J.B. and Bruns, W. (1988) A short synthesis of serricornin, (4S,6S,7S)-4-6-dimethyl-7-hydroxy-3-nonane, the sex pheromone of the tobacco beetle *Lasioderma serricorne* F., from D-glucose. *Carbohyd Res* **174**: 341–348.

285 Reed, W.D. (1935) Notes on the distribution of cured tobacco insects in the Near East *Proc Ent Soc Wash* **37**: 42–48.

286 Reed, W.D. and Vinzant, J.P. (1942) Control of insects attacking stored tobacco and tobacco products. *U.S. Dept Agric, Circ no.* 635.

287 Reed, W.D., Morril Jr., A.W. and Livingstone, E.M. (1934) Trapping experiments for the control of the cigarette beetle – *U.S. Dept. Agric, Circ no.* 356, 13pp.

288 Reif, H.F. (1985) New formula speeds up fumigation. *Tob Rep* Sept.

289 Retief, E. and Nicholas, A. (1988) The cigarette beetle *Lasioderma serricorne* (F.)(Coleoptera: Anobiidae): a serious herbarium pest *Bothalia* **18**:97–99 (in Edwards *et al*, 1990).

290 Richards O.W. (1952) A new species of Bethylidae (Hymenoptera) from Palestine. *Ann Mag Nat Hist* **5**: 409–410.

291 Richards, O. W. and Waloff, N. (1946) The study of a population of *Ephestia elutella* Hübner (Lep., PHYCITIDAE) living on bulk grain. *Trans R ent Soc Lond* **97**: 253–298.

292 Richards, O.W. and Herford, G.V.B. (1930) Insects found associated with cacao, spices, and dried fruits in London warehouses. *Ann Appl Biol* **17**: 367–395.

293 Richards, O.W. and Thomson, W.S. (1932) A contribution to the study of the genera *Ephestia* Gn (including *Strymax* Dyar), and *Plodia* Gn (Lep., Phycitidae), with notes on parasites of the larvae. *Trans R ent Soc Lond* **80**: 169–250.

294 Rizk, G.N., Al-Badry, E.A. and Hafez, S.M. (1980) Biological studies of the predatory mite *Acaropsis solers* attacking stored product pests. *Mesopotamia J Agric* **15**: 203–222.

295 Rizk, G.N., Mostafa, S.A.S., Shaaban, A.M. and Choniem, H.G. (1980a) The population density of the cigarette beetle, *Lasioderma serricorne* (F.) on two types of tobacco in Egypt. *Z ang Ent* **90**: 180–183.

296 Ross, D.H. (1990) Methoprene: Its Development and Use for Tobacco Protection. *Tob Int* 12–18, September 15.

297 Runner, G.A. (1919) The Tobacco Beetle: an important pest in tobacco products. *Bull U.S. Dept Agric* no 737: 49–51.

298 Runner, G.A. (1916) Effect of Röntgen rays on the tobacco, or cigarette, beetle and the results of experiments with a new form of Röntgen tube. *J Agr Res* **6**: 383–388.

299 Runner, G.A. (1917) The tobacco beetle and how to prevent damage by it. *US Department of Agriculture Farm Bulletin* **846**, 3–22.

300 Samfield, M. and Brock, B.A. (1958) Bulk thermal conductivity of tobacco. *Tob Sci* **2**: 49–50.

301 Samuel, R., Prabhu, V.K.K. and Narayanan, C.S. (1984) Influence of spice essential oil on the life history of *Lasioderma serricorne* (F.) *Entomon* **9**: 209–215.

302 Sannino, L., Sorrentino, C. and Balbiani, A. (1983) Prime indagini sulla struttura del corion delle ouva di *Lasioderma serricorne* F. (First surveys on the chorion structure of *Lasioderma serricorne* F. eggs). *Ann Inst Sperimentale Tob* **10**: 95–101.

303 Saxena, S.C. and Gohain, R. (1976) Susceptibility to insecticides of *Lasioderma serricorne* (F.) (Coleoptera Anobiidae) grown on vitamin deficient diets. *Archives Int de Phys et de Biochimie* **84**: 53–61.

304 Schaefer, C.H. and Dupras, E.F. (1973) Insect developmental inhibitors 4. Persistence of ZR-515 in water. *J Econ Ent* **66**: 923–925.

305 Schooley, D.A., Bergot, B.J., Dunham, L.L. and Siddall, J.B. (1975) Environmental degradation of the insect growth regulator methoprene (Isopropyl (2E,4E)-11-methoxy-3,7,11-trimethyl-2,4-dodecadienoate) II. Metabolism by aquatic microorganisms. *J Agr Food Chem* **23**: 293-298.

306 Schulten, G.G.M. and Kuyken, W. (1967) Warehouse moths and some tropical products stored in the Netherlands. *Meded Rijksfac Landbouwwetensch Gent* **32**: 584–592.

307 Self, L.S., Guthrie, F.E. and Hodgson, E. (1964) Metabolism of Nicotine by tobacco feeding insects. *Nature* **204**: 300–301.

308 Shaheen, D.G., Bragg, L.A., Stanovick, R.P. and Sullivan, J.B. (1983) Phosphine detector tubes as passive dosimeters for use with metal phosphide fumigants. *Tob Int* **185**: 36–39.

309 Shaw, Q.A.Q. and Hussain, Z. (1990) Control of Cigarette Beetle, *Lasioderma serricorne* (F) (Coleoptera): Anobiidae) With Methoprene. *Pak Tobacco* **14** (1) 19–21.

310 Shemshedini, L. and Wilson, T.G. (1990) Resistance to juvenile hormone and an insect growth regulator in *Drosophila* is associated with an altered cytosolic juvenile hormone-binding protein. *Proc Matl Acad Sci USA* **87**: 2072–2076.

311 Shen, S.K. and Dowd, D.F. (1990) Xenobiotic induction of esterases in cultures of the yeast-like symbiont from the cigarette beetle. *Ent Exp Applic* **52**: 179–184.

312 Shen, S.K. and Dowd, P.F. (1991) 1-napthyl acetate esterase activity from cultures of the symbiont yeast of the cigarette beetle (Coleoptera: Anobiidae). *J Econ Ent* **84**: 402–407.

313 Shew, H.D. and Lucas, G.B. (1991) *Compendium of tobacco diseases*. APS Press, St. Paul, USA, 68pp.

314 Silvestri, F. (1912) Contribuzioni alla conoscenza degli Insetti dannesi e dei loro simbionti. *Boll Lab Zool Portici* **6**: 246–307

315 Sine, C. (1993) (Eds) *Farm Chemicals Handbook* 1993. Meister Publ. Co. Willoughby, USA.

316 Singh, P. (1977) *Artificial diets for insects, mites and spiders*. IFI/PLENUM, New York British Lending Library Div. no. 77/12298.

317 Singh, S., Dilawari, J.B., Vashist, R., Malhotra, H.S. and Sharma, B.K. (1985) Aluminium Phosphine Ingestion. *British Med J* **290**.

318 Sivik, F.P., Tenhet, J.N. and Delamar, C.D. (1957) An ecological study of the cigarette beetle in tobacco storage warehouses. *J Econ Ent* **50**: 310–316.

319 Smee, C (1944) Report of the entomologist (Nyasaland) (1944) 5pp (referred to by Howe, 1957).

320 Smit, H. (1976) The control of *Lasioderma serricorne* and *Ephestia elutella* in the cigarette industry. *Beit zur Tabak* **8**: 395–397.

321 Soderstrom, E.L. (1970) Effectiveness of green electroluminescent lamps for attracting stored-product insects. *J Econ Ent* **63**, 726–731.
322 Soderstrom, E.L. and Lovitt, A.E. (1968) Chambers for collecting and observing single pupae of three stored product moths. *J Econ Ent* **61**: 1468–1469.
323 Soemartaputra, M.H., Haryadi, R.S., Rahayu, A., Kardha, S., Purwanto, Z.I., Chosdu, R. (1985) Radiation disinfestation of tobacco bales and coffee beans. In: *Food irradiation processing*. International Atomic Energy Agency, Vienna, 170 pp.
324 Stahl, E., Rau, G., Adolphi, H. (1985) Entwesung von Drogen durch Kohlendioxid-Druckbehandlung (PEX-Verfahren) (Disinfestation of drugs by pressure treatment with carbon dioxide (PEX procedure)). *Pharm Ind* **47**: 528–530.
325 Staiger, L.E., Quistad, G.B. and Schooley, D.A. (1983) Stability of (5-^{14}C) methoprene on cigarette filler tobacco in glass jars and bright leaf tobacco in simulated hogsheads. *J Econ Ent* **76**: 999–1001.
326 Stamatinis, N.C. (1935) The enemies of tobacco in warehouses, *Ephestia elutella* Hb. and *Lasioderma serricorne* Fab. The biology and measures for their control. (*In Greek with English summary) Commun Tob Inst Greece*, no 4, 65pp.
327 Steffan, J.R. (1982) L'entomofaune de la momie de Ramese II. *Annales de la Soc Ent de France* **18**:531–537.
328 Stratil, H.H., Reichmuth, Ch. (1984) Kühllagerung von Süsswaren zur Mottenbekämpfung. *Süsswaren* **7–8/84**: 332–334.
329 Strong, R.G. and Diekman, J. (1973) Comparative Effectiveness of Fifteen Insect Growth Regulators Against Several Pests of Stored Products. *J Econ Ent* **66**: 1167–1173.
330 Strong, R.G., Partida, G.J. and Warner, D.N. (1968) Rearing stored-product insects for laboratory studies: six species of moths. *J Econ Ent* **61**: 1237–1249.
331 Subramanyam, BH. and Cutkamp, L.K. (1987) Influence of posttreatment temperature on toxicity of pyrethroids to five species of stored-product insects. *J Econ Ent* **80**: 9–13.
332 Sullivan, J.B. (1985). Studies show Degesch fumigant effective in cold weather. *Tob Int* May: 8–12.
333 Swingle, M.C. (1938) Low temperature as a possible means of controlling the cigarette beetle in stored tobacco. *USDA, Circular* No. 462.
334 Takahashi, F. (1973) Sex pheromones; are they really species specific? *Mem Coll Agric, Kyoto Univ* No. **104**: 13–21.
335 Takaoka, I. and Naito, T. (1968) Control of Stored Tobacco Insect Pests in Japan. *Photoxin® Tech Conf* 15pp., (File copy only).
336 Tallafus, O., Hiller, G. (1991) Kaltentwesung von Lebensmitteln (Cold Disinfestation of Foodstuffs). *ZFL* **42**: 476–480.
337 Taylor, R.W.D. and Halliday, D. (1986) The Geographical Spread of Resistance to Phosphine by Coleopterous Pests of Stored Products. *The 1986 British Crop Protection Conf* paper # 4C-16 pp. 607–613.
338 Tenhet, J.N. (1959) Pyrethrum Mists and Aerosols for Control of Insects in Tobacco Warehouses. *Marketing Res Rep No. 334, USDA*. 21 pp.
339 Tenhet, J.N. and Bare, C.O. (1951) Control of insects in stored and manufactured tobacco. *U.S. Dept Agric, Circ* no. 869.
340 Tenhet, J.N., Bare, C.O. (1946) Redrying of tobacco and its effect on insect infestation. *J Econ Ent* **39**: 607–609.

341 Tenhet, J.N., Bare, C.O., Childs, D.P. (1957) Cold storage and cool storage of tobacco to control the cigarette beetle. *Tob Sci* **1**: 169–174.
342 Tenhet, J.N. (1954): cited from: Expert describes control of insects attacking leaf. *U.S. Tob J* **162**: 5, 24–25.
343 Theobald, F.B. (1911) Report on the economic zoology, (1911). Animals injurious to man's buildings, furniture, stores and food. *J S.E. Agric Coll Wye* **20**: 226–237.
344 Thompson, J.V. and Fletcher, L.W. (1972) A pathogenic strain of *Bacillus cereus* isolated from the cigarette beetle, *Lasioderma serricorne* **20**: 341–350.
345 Tilton, E.E., Burkholder, W.E., Cogburn, R.R. (1966) Effects of gamma radiation on *Rhyzopertha dominica, Sitophilus oryzae, Tribolium confusum,* and *Lasioderma serricorne. J Econ Ent* **59**: 1363–1368.
346 Tobin, E.N. and Smith, L.W. (1971). Note on the mating behaviour of the cigarette beetle *(Lasioderma serricorne* (F): Anobiidae). *Ent News* **82**: 23–25.
347 Todorovski, V.B. and Vasilev, S.L. (1974) Insecticidal Value of Phosphine in the Control of Stored Tobacco Insect Pests. *Proceedings of the 1st International Working Conference on Stored-product Protection,* Savannah GA, USA. October 7–11, pp. 679–685.
348 Tungul, A. (1991) Micronuclei induction by dichlorvos in mouse skin. *Mutagenesis* **6**: 405–408.
349 Underwood, J.G. (1972) The Radiochemical Analysis of Phosphine Residues in Fumigated Tobacco. *Tob Sci* **16**: 123–126.
350 U.S.D.A. (1972) Stored Tobacco Insects – Biology and Control. *U.S. Dept Agric Handbook* no 233.
351 U.S.D.A. (1980) Guidelines for the control of insect and mite pests of foods, fibres, feeds, ornamentals, livestock, households, forests, and forest products. *Agric Handbook* no. 571.
352 U.S.D.A./A.P.H.I.S. (1985) *Plant Protection & Quarantine-Treatment Manual,* May 29.
353 Ustinov, A.A. (1932) A review of the pests of tobacco in Abkhazia observed in 1931. (Russian) Sukhum, *Abkhazsk tabach zonal'n Sta.*
354 Van Emden, F. (1929) Ueber die rolle der Feuchtigkeit im leben der Speicherschadlinge. *Anz Schadlingsk* **5**: 58–60.
355 Vasilev, L. (1963) Pratical Application of Phostoxin® in the Control of Stored Tobacco Moth (*Ephestia elutella*). *Tutun/Tobacco* **13**: 227–231.
356 Vincent, L.E. and Lindgren, D.L. (1977) Toxicity of Methyl Bromide and Phosphine at Various Temperatures and Exposure Periods to the Metamorphic Stages of Lasioderma Serricorne. *J Econ Ent* **70**: 497–500.
357 Voges, E. (1984) *Voges Tobacco Encyclopedia* ISBN 3–920615–07–7.
358 Waloff, N. (1948) Development of *Ephestia elutella* Hb. (Lep., Phycitidae) on some natural foods. *Bull Ent Res* **39**: 117–130.
359 Waloff, N. (1949) Observations on larvae of *Ephestia elutella* H_bner (Lep., PHYCITIDAE) during diapause. *Trans R ent Soc Lond* **100**: 147–159.
360 Waloff, N. and Richards, O.W. (1946) Observations on the behaviour of *Ephestia elutella* Hübner (Lep., PHYCITIDAE) breeding on bulk grain. *Trans R ent Soc Lond* **97**: 299–335.
361 Waloff, N., Norris, M.J. and Broadhead, E.C. (1948) Fecundity and longevity of *Ephestia elutella* H_bner (Lep., PHYCITIDAE). *Trans R ent Soc Lond* **99**: 245–268.
362 Ward, W.H., Fletcher, L.W., Childs, D.P. (1982) Spunbonded polyethylene

sheet as a barrier for the cigarette beetle and its effect on chemical and moisture changes in stored tobacco. *Tob Sci* **26**: 30–31.

363 Weiser, J. and Hostounsky, Z. (1971) Host to host transmission of *Nosema plodiae. J Protozool* **18**: 29.

364 Wilbur, D.A. (1983) Magnesium Phosphide. *3rd International Working Conference on Stored-product Entomology* Manhatten, KS, USA, October 23–28, pp. 341–345.

365 Wohlgemuth, R. (1973) Vergleichende Untersuchungen über die Wirkung von Röntgen- und Gammastrahlung (60 Co) auf den Kornkäfer (*Sitophilus granarius*) (Studies on the effect of ionizing radiations on the viability of the adult grain weevil (*Sitophilus granarius*)). *Z angew Ent* **74**: 7–24.

366 Wohlgemuth, R. (1985) Phosphine permeability of various plastic sheets and films. *Bull OEPP* **15**: 59–64.

367 Wohlgemuth, R., Reichmuth, C., Rothert, H. and Bode, E. (1987). The appearance of moths of the genus *Ephestia* and *Plodia* which are harmful to stored products, outside of warehouses and food processing factories in Germany. *Anz Schad, Pflan Umwelt* **60**: 44–51.

368 Wojcik, D.P. (1968) Tests for audible and ultrasonic sound production by stored-product insects. *J Econ Ent* **61**: 1415–1416.

369 Wojcik, D.P. (1969) Monitoring for audible and ultrasonic sound production by stored-product insects during mating. *J Econ Ent* **62, 4**: 937.

370 Yamamoto, R.T. and Fraenkel, G. (1960) The suitability of tobaccos for the growth of the cigarette beetle, *Lasioderma serricorne. J Econ Ent* **53**: 381–384.

371 Yamasaki, M. (1982) Biology of a sanitary injurious bethylid wasp, *Cephalonomia gallicola* (Ashmead) (Hymenoptera, Bethylidae). *Jap J of Sanitary Zool* **33**: 221–226.

372 Yang, S.S. (1992) A rapid method for the determination of methoprene in tobacco by high performance liquid chromatography. *Chroma* **33**.

373 Yokoyama, V.Y. and Mackey, B.E. (1987) Relation of plant protein and suitability of cotton foliage for cigarette beetle (Coleoptera: Anobiidae) growth. *J Econ Ent* **80**: 830–833.

374 Zettler, J.L., Keever, D.W. and Minor, M.F. (1994) Phosphine resistance in the cigarette beetle, *Lasioderma serricorne* (F) (Coleoptera: Anobiidae) from tobacco storages in the Southeastern United States. *Econ Ent* (in press).

375 Zoecon (1980) *Kabat Product Manual*, Section III, P.7.

376 Zoecon (1977) *Technical Bulletin, Kabat Insect Growth Regulator (methoprene) Z5-13-(7–77)* Zoecon Corporation Palo Alto, Calif. 94304.

Index

Printed in the United States
87842LV00003B/349-351/A